Attribution Theory in the Organizational Sciences

Theoretical and Empirical Contributions

A volume in
Advances in Attribution Theory
Series Editor: Mark J. Martinko

Attribution Theory in the Organizational Sciences

Theoretical and Empirical Contributions

by
Mark J. Martinko
Florida State University

INFORMATION AGE
PUBLISHING

80 Mason Street • Greenwich, Connecticut 06830 • www.infoagepub.com

Library of Congress Cataloging-in-Publication Data

Florida State International Symposium on Attribution Theory (2nd : 2004)
  Attribution theory in the organizational sciences : theoretical and
empirical contributions / edited by Mark J. Martinko.
     p. cm. – (Advances in attribution theory)
  "Product of the 2nd Florida State International Symposium   on
Attribution Theory held in February 2004"–Frwd.
  Includes bibliographical references and index.
  ISBN 1-59311-125-8 (pbk.) – ISBN 1-59311-126-6 (hardcover)
  1. Attribution (Social psychology)–Congresses. 2. Organizational
behavior–Congresses. I. Martinko, Mark J. II. Title. III. Series.
  HM1076.F56 2004
  302.3'5–dc22
                                        2004014524

Printed in the United States of America

*This book is dedicated to the legacy of Fritz Heider
and the many scholars who continue to work on and
extend his perspective of human behavior.*

CONTENTS

FOREWORD

This book is the product of the 2nd Florida State International Symposium
on Attribution Theory held in February 2004. This first symposium was
held 10 years earlier in 1994 and received considerable attention from the
academic community. The book from that conference, *Attribution Theory:
An Organizational Perspective*, was cited by Shafritz and Ott's *Classics of Organization Theory* (2000) as one of the two most significant contributions to
organization theory in 1995. We are obviously hopeful that this book will
have a similar impact.

The purpose of the second symposium as well as that of this book is to
provide an in-depth forum for the discussion, integration, dissemination,
and development of both research and theory describing the nature, role,
and contribution of attribution theories to understanding the dynamics of
organizational behaviors.

Implicit in the above objectives is the notion that attributional processes
are a fundamental but sometimes ignored part of the paradigm of achievement-related behaviors in organizations. Thus, attribution processes affect
and are affected by a wide variety of variables in organizational settings.
The research and conceptual articles that make up this volume reflect that
reality, ranging from micro to macro behavioral topics. Examples of the
range of topics include how attributions are related to moral behavior, the
pacing and spacing of goals, emotional states such as negative affectivity
and burnout, traits associated with core self-evaluations, workplace aggression, leader and member relations, conflict, joint ventures, and leaders'
responses to crisis situations. We recognize that many of these contributions appear to stretch the bounds of attribution theory and we are hopeful that these adaptations prove to be productive both in terms of
providing better explanations for organizational behaviors but also contribute to the continuing evolution of attribution theory.

I am grateful to all of the people who have supported this effort. All of
the chapters were reviewed by at least three scholars who voluntarily con-

Attribution Theory in the Organizational Sciences, pages ix–x
Copyright © 2004 by Information Age Publishing
All rights of reproduction in any form reserved.

tributed their time and feedback. Their names appear in the list of reviewers below. A deep sense of appreciation is also extended to all of the authors who contributed their chapters to both the conference and this book. There is now a proliferation of conferences that vie for scholars' attentions. This competition, coupled with somewhat limited budgets, makes researchers' choices difficult, so their participation is doubly appreciated. The contributions of Bernard Weiner are likewise recognized. His participation as the keynote speaker and contributor was invaluable. My former and current doctoral students also made significant contributions to the conference and include Constance Campbell, Scott Douglas, Bill Gardner, Paul Harvey, Russell Kent, Sherry Moss, and Neal Thomson. Scott Douglas was particularly helpful and served as a sounding board throughout the development of the conference. Finally, I would like to thank Dean Stith and Jerry Ferris of Florida State University for their financial support of this effort.

Mark J. Martinko
Bank of America Professor of Management
Florida State University

REVIEW BOARD

Neal Ashkanasey
Hector Betancourt
Constance Campbell
Scott C. Douglas
William L. Gardner
Michael Gundlach
Gerald Ferris
Jerald Greenberg

Ken Harris
J. G. Hunt
Russell Kent
Jeff LePine
Robert C. Liden
Robert G. Lord
Fred Luthans
Sherry E. Moss
Robert Zmud

Ann O'Leary-Kelley
Pamela Perrewe
Sandra Robinson
Elizabeth Rozelle
Alex Stajkovic
Neal Thomson
Bernard Weiner
Kelley Zellars

## CHAPTER 1

ATTRIBUTION THEORY AND ORGANIZATIONAL PSYCHOLOGY

An Introduction and Overview

Bernard Weiner
University of California, Los Angeles

A theory and the domain to which it is applied are similar to a match or pairing between two marital partners. In some instances, the couple seems perfect and surely the marriage will last. Of course, sometimes we are disappointed that a relationship "made in heaven" does not work out, but usually those pairings succeed and are envied by others. Evolutionary theory and drive theory are good matches to explain the behavior of hungry rats as they search for food. After all, nourishment is necessary for survival, which is the bedrock of these theories. It is probably a relationship that will last, and we admire the ability of the conceptions to shed light on behavior. Conversely, attribution theory is likely to be a poor mate for this viscerogenic behavior. Using attribution theory to predict food consumption would end in a quick divorce.

For organizational behavior, it is unlikely that evolutionary or drive theory will provide a permanent partner. One can think of the survival of organizations, and a need for profits, but these are not rich conceptual

Attribution Theory in the Organizational Sciences, pages 1–3

frameworks for this setting. On the other hand, attribution theory appears (at least to me) to be a loving and caring spouse. This is because in organizational settings individuals seek to fulfill personal goals, while at the same time others within that setting react to the person. That is, there are both intrapersonal and interpersonal motivational issues. As it happens, attribution theory has associated with it intrapersonal and interpersonal concerns and has provided the constructs for the development of a variety of theories to account for both. What good fortune!

This book, organized and edited by Mark Martinko, brings together researchers and practitioners from around the United States and from the international community who apply attribution theory in organizational contexts. Many of the contributors first came together 10 years earlier for a conference held at Florida State University in Tallahassee, which resulted in a very well received book. In a fortuitous occurrence, this conference was scheduled on February 20; Fritz Heider, the "founder" of attribution theory, was born on February 19 (in 1896). He would have been very pleased to hear these papers and proud of the extension of attribution theory into organizational contexts. The chapters in this book are on the cutting edge of the attribution/organization intersection.

The contributors varied greatly in their research focus. One expressed interest in crisis management; another in international conflict; a third in the cycle of work; burnout was the topic of a fourth contributor; self-evaluations concerned others; individual differences in perceptions of the causes of success and failure of the self and/or others was addressed; and on and on. Remarkably, in spite of this diversity, all either solely or in conjunction with other approaches called upon attribution theory to provide the foundation for their thinking and research. Furthermore, they not only used this conceptual scaffold, but often extended it in new directions, generating hypotheses not previously formulated by attribution theorists.

The approach of organizational psychologists provides an important supplement, or complement, to that of experimental social psychologists associated with attribution theory. The work presented here (with some exceptions) can be described as:

1. Correlational rather than manipulating experimental variables
2. Focused on individual differences
3. Recognizing the role of affect in human behavior
4. Applying attribution theory to both the perception of self and others
5. Accepting that attributions have underlying properties of locus, stability, and control, with these properties linked with disparate psychological consequences.

Given these foci, the presenters faced a set of core problems. They include:

1. What is an attribution, and how does it differ from a social inference and from an attribute?
2. How should attributions be assessed? Should one examine specific attributions such as ability and effort as causes of success and failure, or the underlying locus, stability, and controllability properties that characterize these causes?
3. How should affects be measured? Are self-reports of immediate affective states valid indicators of feelings?
4. If an attributional style is identified, is it consistent across time and situations in a manner similar to other traits? And what are the relative contributions of states and traits, or individual differences and situational variables, to behavior?

Thus, very basic issues in attribution theory are confronted in this book. Of course, there are varying degrees of definitive closure, depending on the problem addressed. The half-full perspective highlights the great strides and progress evident in the book. To whet the appetites of browsers of this introduction, among the attribution-derived hypotheses and/or empirical findings are: the action cycle of goals in part depends on whether deadlines have been met and the perceived reasons for not fulfilling these deadlines or meeting them early; personality variables related to attributional style influence work attitudes; attributions for burnout determine its consequences; self-evaluations influence self-serving causal biases; attributional style is a predictor of aggression; conflict resolution depends on attributions for the problem and emotions aroused by causal beliefs; social attributional style predicts judgments of others; followers can infer the intentions of leaders and discriminate selfish from organizational goals; empathy is an important component of conflict resolution along with attributional beliefs; the success or failure of international ventures is in part determined by other-blame tendencies that differ between cultures; and on and on. This dazzling array will convince the most skeptical reader of the richness of the attribution/organization nexus.

For others who might regard the cup half-empty rather than half-full, this more pessimistic perspective also has a positive note. It suggests a third conference is needed. Knowing the editor, I believe that will occur, for the resolution of old problems generates new ones, and some of the original problems noted previously (e.g., how to measure attributions and affects) are exceedingly complex and will persist. Hence, there is much to look forward to in the years ahead.

CHAPTER 2

SOCIAL MOTIVATION AND MORAL EMOTIONS

An Attribution Perspective

Bernard Weiner
University of California, Los Angeles

In my initial work with attribution theory I was very much guided by Fritz Heider (1958) and Harold Kelley (1967) and their metaphor that the person is a scientist, trying to gather information in a reasonably rational way and reaching decisions regarding the causes of behavior. In the past two decades, however, I turned from intrapersonal to interpersonal behavior, and from the metaphor of the person is a scientist to the person is a judge (see Weiner, 1995). In this chapter, I focus on my most recent work regarding interpersonal behavior, first discussing the metaphors driving my attributional approach and then recent meta-analyses of help-giving and aggression that test an attribution–emotion–action conception that has been developed.

Following this, I shift from theory testing to a very recent interest concerning moral emotions and particularly to some neglected emotions in achievement contexts including admiration and envy. After introducing that topic, I present research regarding personality inferences of arrogance and modesty because these inferred characteristics are related to moral emotions. These are the types of emotions and inferences that are

Attribution Theory in the Organizational Sciences, pages 5–24
Copyright © 2004 by Information Age Publishing
All rights of reproduction in any form reserved.

likely to be prevalent in organizational contexts and should profoundly affect output, worker satisfaction, and indeed the survival of that setting.

THE INTERPERSONAL THEORY

Guiding Metaphors

My theoretical approach is guided by two metaphors. In a metaphor, we take what is known about one object and use it to explain the unknown. For example, if I call a person a shark, then that calls attention to the characteristics of a shark, such as cunningness, aggressiveness, and so on, that might shed light on this individual. Or we might call the person a rock, which makes salient her resilience under stress and ability to face failure. Using these metaphors, we then see the person through a new lens and are alert to find associated attributes. Note that there can be multiple metaphors operating at a single time; that is, metaphors can stand side by side in the explanation of behavior.

In the field of human motivation, we need metaphors that help us understand the unknown. One metaphor that has proven useful, albeit is now outdated, construes the person as a machine, as the behaviorists did (and perhaps do). This resulted in a spate of research regarding input–output relations, without the interventions of cognitions and emotions. A contrasting current metaphor regards the person as a computer, with all the implications of that description (see Weiner, 1991).

One metaphor I use that guides my view of interpersonal behavior is we are judges, determining if others are good or evil, right or wrong, moral or immoral. A second related metaphor I find heuristic is life is a courtroom where dramas related to moral issues are played out. Consider, for example, the following scenario. You arrange to meet a friend to go to the movies together. Imagine you will meet in front of the theater at 8:00. You arrive on time at 8:00 but your friend is not there. And he or she is not there at 8:15 or 8:30. By this time you are likely to be flooded with a variety of emotions. Perhaps you fear something has happened, like a car accident, remembering he is a rather bad driver. Or, you may be concerned there was a misunderstanding, since this movie is playing at two different theaters. And you may be experiencing anger, knowing this person often is late.

Then after 8:30 your friend arrives. Being a good attribution theorist, you ask "Why are you so late?" That is the attribution of causality question, typically given following unexpected and negative events. Imagine the person says, "I walked very slowly getting here and took the long route." Now, most criminals do not self-incriminate, but this is what the person says. You respond with anger, holding the other responsible for making you wait and missing the movie. You then sentence the person: "I will never meet you again and this ends our friendship." But the transgressor then adds: "A terrible thing just happened to me. My sister was in a bad car accident. I needed the time to get myself together. I did not want to ruin our date." This is a mitigating circumstance and the prisoner is likely to be paroled, so that friendship is renewed.

Here we see life as a courtroom; we act as judges and moral life and social life are intertwined as we apply moral rules to everyday social contexts. That is, the courtroom metaphor suggests social life is guided by the same moral and religious principles that provided the bases for the law. This is the lens I use to understand social motivation and the kinds of observations I attempt to capture. This metaphor is grounded in law and theology, so these areas serve as the foundation for my approach to motivation, rather than, say, the neurosciences and the metaphor the person is a machine.

An Attributional Theory of Social Motivation and Justice

A metaphor in and of itself is not an explanation but rather provides the foundation for the next scientific step, which is to build a theory capturing the ideas in the metaphor. Let me then present a theory that formally captures the story I just told, along with research domains in which the theory has been tested (for a full review, see Weiner, 1995). This theory is shown in Figure 2.1.

In the lefthand column of Figure 2.1 it can be seen an event has occurred related to five domains of study I have focused upon: achievement evaluation, stigmatization, help-giving, compliance with a request to transgress, and aggression. The top half of the table includes those instances in which the person is held responsible given a negative outcome. Achievement failure, shown in the top row, is ascribed by an observer (not necessarily by the actor as well) to a lack of effort. In this chapter I do not examine the antecedents that give rise to particular causal beliefs, so merely accept lack of effort is the perceiver's explanation for the outcome. Lack of effort is regarded a controllable cause—it could have been otherwise. If there are no mitigating circumstances, then the person is held responsible for the failure. For involved parties, such as the teacher or parent, this appraisal gives rise to anger. Imagine, for example, your reaction to your child when he or she is not studying and doing poorly in school, or when a player on your team or the team of your city is not putting forth effort during a game. Anger, in turn, elicits some form of negative or punitive reaction.

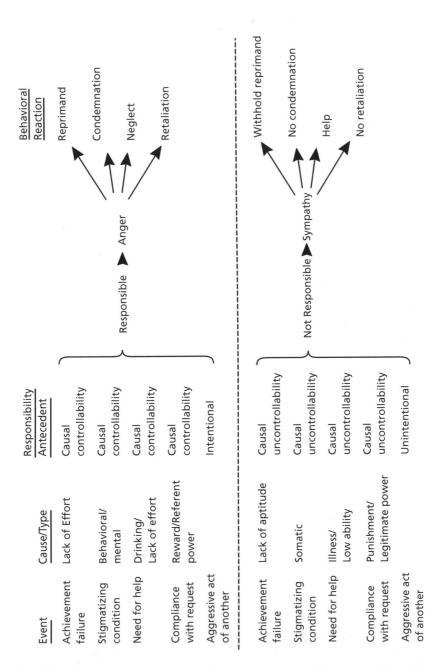

Figure 2.1. An attributional theory of social motivation.

Preceding to the bottom half of the figure and the achievement context, here failure is ascribed to lack of aptitude. Lack of aptitude is not perceived as under volitional control—it is not subject to willful change. Hence, the failing person is not held responsible for the outcome and the reaction is sympathy rather than anger Imagine your feelings and actions toward the academic failure of a mentally handicapped person, or to a failure in athletics of someone physically handicapped. The response to this is prosocial, such as providing support and withholding punishment.

The other rows in the figure capture this process in other domains. For example, under stigmatization, imagine another obese because of overeating (which would be in the top half of the table) as opposed to a thyroid problem (located in the bottom half of the table); has a heart problem because of lack of exercise as opposed to a genetic shortcoming; or has AIDS because of promiscuous sexual activity as opposed to contracting AIDS during a blood transfusion. Here again the same sequence is presumed: from a particular causal belief and causal property (controllability) to an inference about the person (responsibility), and then to affects (anger and sympathy) and finally to behavior (anti- or prosocial). For example, AIDS because of promiscuous sexual behavior is controllable, the person is regarded responsible, and this gives rise to anger and antisocial responses, whereas AIDS due to a blood transfusion is not controllable by the victim, who is judged not responsible, and sympathy and prosocial reactions are evoked. The motivation sequence, then, is from thoughts to affect to action. I regard this as akin to a "deep structure" for the field of motivation (Ickes, 1996). The same sequence follows in the help-giving, compliance, and aggression domains, which are examined quite soon.

It is important to note I am not presenting a theory of reactions to the stigmatized, or of helping, or of aggression. Rather, one principle is used to explain comparable findings across motivation domains. I believe this is the path to construct a general theory of motivation and it is not possible to have a complete theory of achievement, or helping, or aggression because of the overdetermination of these actions. But this takes me far from the focus of this chapter and ends the introduction.

Helping and Aggression: A Meta-Analytic Review

I now present meta-analytic work regarding helping and aggression that tests the theory just presented. The helping research includes help for stigmatized groups, so this topic is also incorporated into the chapter. Help-giving and aggression have proven to be the best social domains to test the theory in its entirety because these areas have generated the most complete research designs. You will soon see that the results are somewhat surprising.

To test the full theory, along with three associates (Rudoph, Roesch, Greitemeyer, & Weiner, 2004), we performed a meta-analysis of the pertinent investigations. In this procedure, one uses all the archival data that can be found. We started with the PsychLit database using a variety of keywords I won't elaborate here. We then traced all references in the articles we uncovered, as well as contacting the authors regarding other pertinent publications. We searched nearly 2,000 articles. To be included in our analysis, the investigation had to have at least one attribution variable (e.g., controllability, responsibility), and one emotion (e.g., anger, sympathy) or one behavioral variable (e.g., help or social support), as well as the first-order correlations between the variables.

As a result of this search process, 39 studies were obtained that met our criteria in regard to helping. The number of participants in these studies was nearly 8,000, so there was an average of about 200 participants per study. For aggression, there were 25 studies with about 4,500 respondents, so again around 200 participants per study. Thus, we have more than 12,000 respondents, a substantial number of subjects in psychological research.

To review quickly what was just presented, the theories in regard to help-giving and aggression are identical, as reviewed in Figure 2.2. Here again we see that there is a precipitating event (a person in need; a hostile act). These give rise to attributions for the event, including classification of the

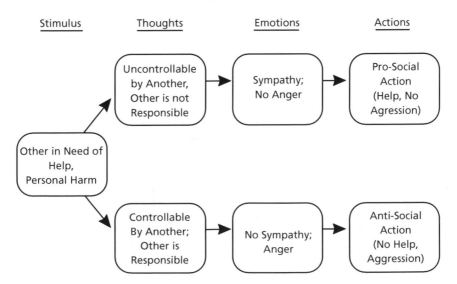

Figure 2.2.   Help-giving and aggression from an attributional perspective.

cause as controllable or not and the person as responsible or not. These thoughts, in turn, give rise to emotions of anger (given responsibility) and sympathy (given no responsibility), which then are proximally related to anti- or pro-social actions (i.e., help vs. neglect and retaliate vs. forgive).

These two domains are rarely incorporated within the same conception. Evolutionary psychologists can embrace both and so can learning theorists who follow a modeling approach. Otherwise, these are quite separate areas of study examining, for example, the effects of diffusion of responsibility on help-giving, or the effects of temperature level on aggression.

We tested a number of different models in our analyses, including the four shown in Figure 2.3.

1. Model 1 includes not only a direct path from emotion to action, but also one from thinking to doing, so attributions have both a direct and an indirect effect on action.
2. Model 2 is the one championed thus far, with thinking distal and emotions proximal to action.
3. Models 3 and 4 include only one of the two affects specified in the model. These models were tested because in some studies only one of the affects was assessed. Therefore, to increase the number of participants being examined, we also considered models that contained only one of the emotional mediators.

We also tested other models that began with emotions. None of these proved significant and I will not discuss them further.

Table 2.1 shows the raw weighted correlations between the variables, along with the number of subjects for each correlation, for the helping data. The table reveals controllability is negatively related to sympathy (r = -.45), positively to anger (r = .52), and negatively to helping (r = -.25), with reasonably high correlations. That is, the more controllable the need (e.g., financial need because of laziness as opposed to the lack of job opportunities, or AIDS because of promiscuous sexual behavior as opposed to a blood transfusion), the less one receives sympathy, the more others are

Table 2.1. Overall Weighted Correlations among Controllability, Sympathy, Anger, and Behavior (Help-Giving) (from Rudolph, Roesch, Greitemeyer, & Weiner, in press)

	Sympathy		Anger		Behavior	
	r	N	r	N	r	N
Controllability	-.45	7416	.52	7140	-.25	6840
Sympathy			-.39	5484	.42	7382
Anger					-.24	6800

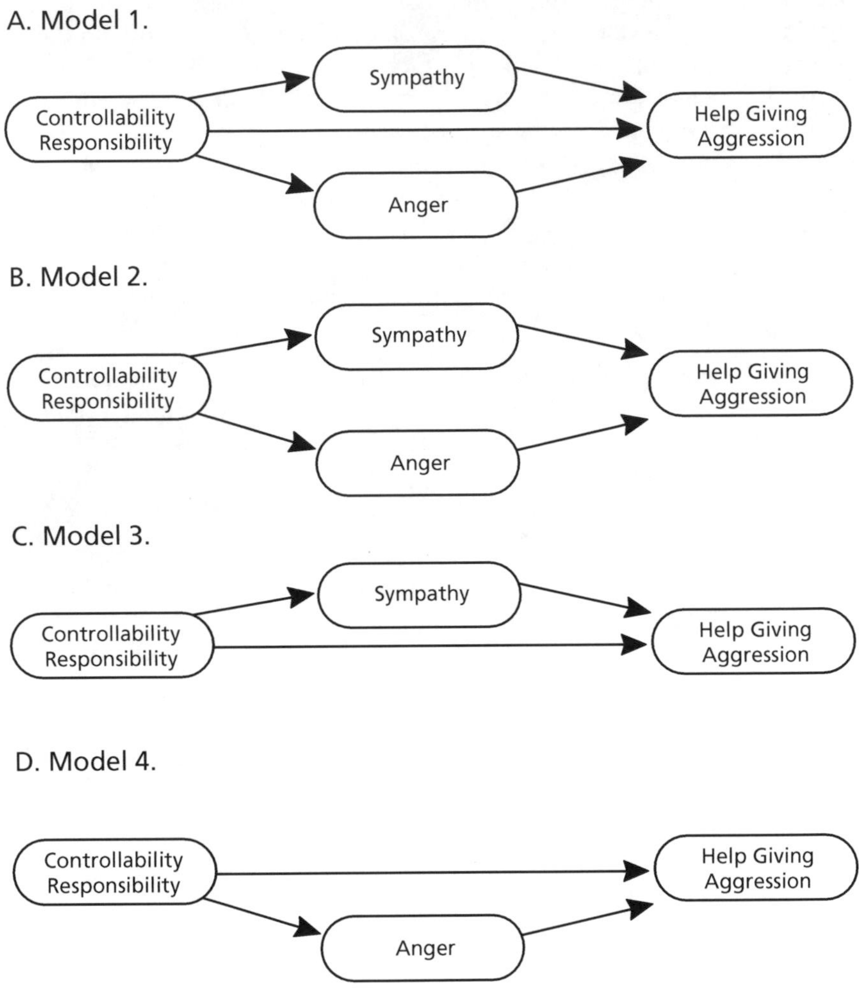

A. Model 1.

B. Model 2.

C. Model 3.

D. Model 4.

Figure 2.3.   Four models of help-giving and aggression from an attributional perspective.

angry, and the less the likelihood of receiving help. It also is evident that sympathy positively predicts help-giving (r = .42), and anger relates negatively to help (r = –.24). Regarding the three predictors of help (control and the two attributions), sympathy clearly is the strongest.

The next table, Table 2.2, shows the path models. Here I concentrate only on comparing the fit of Model 1, which has the direct path from thinking to help, with the fit of Model 2, without such a path. It is clear that the associations in the two models are nearly identical and control is barely related to help-giving (β = –.05, but significant because of the large number of subjects).

Table 2.2.   Path Coefficients for Help-Giving Models (from Rudolph et al., in press)

	Model 1	Model 2	Model 3	Model 4
Help-Giving Models				
Control–Sympathy	−.45*	−.45*	−.45*	—
Control–Anger	.52*	.52*	—	.52*
Control–Help–Giving	−.05*	—	−.08*	−.15*
Sympathy–Help–Giving	.37*	.39*	.39*	—
Anger–Help–Giving	−.07*	−.09*	—	−.17*

* $p < .05$

Both models have excellent fits to the data and do not significantly differ from one another. Using the criterion of parsimony, which posits one of the criteria for theory choice is the number of constructs and postulated associations, the model without the direct path from attributions to behavior is considered the better of the two inasmuch as it has fewer linkages. A number of moderator variables also were considered, such as publication status or date and type of sample, none of which proved significant. Of perhaps most interest is a comparison between simulation versus "real" studies because attribution researchers have often been accused of using only "imaginary situation" research, which could promote controlled processes and the logical connections that have been posited. However, the relations in these two types of investigations are quite similar, as can be seen in Table 2.3. In sum, the full pattern of data is very consistent with the theory and supports the proposed thinking–feeling–acting motivational sequence.

Table 2.3.   Path Coefficients for Help-Giving (Model 1) for the Moderator Type of Investigation (from Rudolph et al., in press)

	Simulated Level	Real Event Level
Control–Sympathy	−.47*	−.37*
Control–Anger	.54*	.45*
Control–Help–Giving	−.07*	.01*
Sympathy–Help–Giving	.40*	.28*
Anger–Help–Giving	−.04*	−.12*

* $p < .05$

What about the data related to aggression or, more accurately, aggressive retaliation? In Table 2.4 one can find the raw correlations for aggression. Table 2.4 reveals the relations again are very strong. The more controllable the aggression (e.g., purposive rather than accidental), the

Table 2.4. Overall Weighted Correlations among Controllability, Sympathy, Anger, and Behavior (Aggression) (from Rudolph, et al., in press)

	Sympathy		Anger		Behavior	
	r	N	r	N	r	N
Controllability	−.35	2509	.61	4448	.49	3719
Sympathy			−.31	1976	−.44	2377
Anger					.56	3458

less the sympathy (r = −.35), the greater the anger (r = .61), and the more the aggression (r = .49). Regarding the determinants of aggression, all the proposed variables are significant, with anger (r = .56) having the strongest link to action.

Next consider for a moment a comparison of the helping and aggression raw correlations (see Table 2.5). It is of some interest to note the positive emotion of sympathy is more predictive of prosocial behavior (the sympathy–help link is r = .42), whereas the negative emotion of anger is more predictive of antisocial behavior (the anger–aggression association is r = .56). I believe positive emotions predict positive actions more than negative inhibitory emotions, and negative emotions predict antisocial behavior more than positive inhibitory feelings. In addition, the correlations predicting to aggression are higher than those for help-giving, intimating aggression may be more amenable to an attribution analysis than is helping. I will soon return to this issue.

Table 2.5. Overall Weighted Correlations among Controllability, Sympathy, Anger, and Behavior (Help–Giving and Aggression) (from Rudolph, et al., in press)

	Sympathy		Anger		Behavior	
	r	N	r	N	r	N
Help–Giving						
Controllability	−.45	7416	.52	7140	−.25	6840
Sympathy			−.39	5485	.42	7382
Anger					−.24	6800
Aggression						
Controllability	−.35	2509	.61	4448	.49	3719
Sympathy			−.31	1976	−.44	2377
Anger					.56	3458

Now turn back to aggression and to the path coefficients, which are shown in Table 2.6. For aggression, the control-to-behavior link is fairly strong and quite significant (although again not as strong as the emotion-to-behavior links). Both Model 1 and Model 2 again significantly fit the data but in this case the model including the path from thinking to behavior provides a stronger fit to the data. That is, for aggression the less parsimonious model is needed to capture the data. Again the simulation findings were similar to the real data and I will not pursue that issue further.

Table 2.6.   Path Coefficients for Agression Models (from Rudolph et al., in press)

	Model 1	Model 2	Model 3	Model 4
Control–Sympathy	−.35*	−.35*	−.35*	—
Control–Anger	.61*	.61*	—	.61*
Control–Aggression	.17*	—	.38*	.24*
Sympathy–Aggression	−.27*	−.30*	−.31*	—
Anger–Aggression	.38*	.48*	—	.42*

* $p < .05$

Table 2.7 shows the comparison of the helping and aggression path data. The table suggests the findings are stronger for aggression. Indeed,

Table 2.7.   Path Coefficients for Help-Giving and Aggression Models (from Rudolph et al., in press)

	Model 1	Model 2	Model 3	Model 4
Help-Giving Models				
Control–Sympathy	−.45*	−.45*	−.45*	—
Control–Anger	.52*	.52*	—	.52*
Control–Help–Giving	−.05*	—	−.08*	−.15*
Sympathy–Help–Giving	.37*	.39*	.39*	—
Anger–Help–Giving	−.07*	−.09*	—	−.17*
Aggression Models				
Control–Sympathy	−.35*	−.35*	−.35*	—
Control–Anger	.61*	.61*	—	.61*
Control–Aggression	.17*	—	.38*	.24*
Aggression Models				
Sympathy–Aggression	−.27*	−.30*	−.31*	—
Anger–Aggression	.38*	.48*	—	.42*

* $p < .05$

the Model 1 fit for aggression is stronger than either the Model 1 or Model 2 fit for help-giving. In particular, the control-anger and anger-behavior paths for aggression far exceed those for help-giving.

What, then, can be said about the generality of the theory across these two phenotypically diverse settings? At the most molar level, it can be concluded that behavior is a function of cognition and affect. This perhaps provides one among the many alternatives to the equally broad statement that behavior is a function of the person and the environment. A somewhat more precise formulation of this general law is emotions are proximal in accounting for both help-giving and aggression. However, thoughts are also proximal determinants of aggression. That is, aggression has more immediate attribution or attribution-derived determinants than does helping. Helping immediately involves only the heart while aggression proximally involves both the heart and the head. Hence, it is probably best to elicit help by appealing to "bleeding hearts" but aggression may be prevented by appealing directly to reason as well as to emotion.

Why might this be the case? One can only offer guarded speculation. It may be that most help and social support have relatively minor personal consequences. Lending someone $10 or driving them to the airport does not involve great cost. On the other hand, even minor aggressive retaliation might come with great harm. Thus, having thinking proximal in aggressive contexts is quite functional.

A related finding already mentioned is aggressive retaliation is better suited to an attribution explanation than is help-giving. Again, why might this be the case? It may be that if someone is attacked, attribution-related thoughts regarding intention and purpose immediately come to mind, so one asks: "Did he do this intentionally? What was the purpose?" On the other hand, a number of nonattribution-related questions also are elicited in helping contexts, such as "Is this person a relative of mine?" "How great is the need?" "Will my help really be beneficial?" That is, nonattribution factors may contribute more to the prediction of help-giving than to aggressive retaliation, although, of course, attributions do play a major role in both domains.

This ends my discussion of this attributional theory of social motivation and justice. I now turn to two of the offspring of this theory: the moral emotions and associated personality inferences.

THE MORAL EMOTIONS

While considering the theory and feeling rather satisfied with the data, I was struck with a few shortcomings (I refuse to reveal the many others but realize readers have their own lists). One limitation is only two emotions, sympathy and anger (and related feelings such as, for example, pity and annoyance), are included in the analysis. Surely there are other emotional reactions to, for example, achievement failure. Second, the theory focus is on moral transgressions, or negative events, including achievement failure, stigmatization, need for help, compliance to commit a transgression, and aggression. What about positive events, such as achievement success or giving funds to charity? These elicit emotional reactions from others and the reactions intuitively seem likely to be guided by attributions for the outcomes. These thoughts led me down two initial paths: what might be other so-called moral emotions in addition to anger and sympathy (i.e., emotions linked to appraisals of ought and should and to controllable vs. noncontrollable causes), and more specifically, what are moral emotions in the context of success?

The analysis I present next regarding the identification of moral emotions is not data driven but rather results from my intuitions and is not meant to be exhaustive. What is offered here is a starting point for research generation, so it is presented with modesty and trepidation. Indeed, I will soon present data suggesting some of the ideas to be expressed already have been found to be questionable.

The moral emotions are shown in Table 2.8. The table distinguishes moral emotions on three dimensions: (1) if the emotion is self- vs. other directed (e.g., guilt versus gratitude); (2) if the emotion is ability- versus effort-linked (i.e., if the emotion falls within the top or bottom half of the theory shown in Figure 2.1); and (3) whether the reaction is positive or negative. In Table 2.8 it can be seen that 12 moral emotions have been

Table 2.8.   Classification of the Moral Emotions

		Emotional Target	
		Self	Other
Causal Link	Ability	Shame (−)	Envy (−)
			Scorn (contempt) (−)
			Sympathy (+)
	Effort	Guilt (−)	Admiration (+)
		Regret (−)	Anger (−)
			Gratitude (+)
			Indignation (−)
			Jealousy (−)
			Schadenfreude (−)

identified. For example, envy, scorn, and contempt are linked to ability, are other-directed and negative. If another fails because of low ability, sympathy is not the only possible reaction. Instead, one may regard the other as a lesser person, eliciting scorn and contempt, hence elevating the self and not giving rise to sympathy (see Izard, 1977). I regard scorn and contempt as "immoral emotions" in that the "proper" or "moral" reaction to uncontrollable failure "ought to be" prosocial.

The observation of scorn rather than sympathy given lack of ability as the cause of failure does not invalidate the structure of the proposed theory inasmuch as moderators are possible at each linkage (including the link between attribution and emotion). Given the negative emotion of scorn, antisocial reactions will follow. That is, an attribution–affect–action sequence, which I regard as the "deep structure" for motivation, holds. Envy also has the quality of an immoral reaction to the success of others and will be negatively evaluated by observers (after all, we judge the emotions of others as well as their behaviors).

It also can be seen in Table 2.8 that the majority of moral emotions are effort-linked, directed toward others, and negative. For example, *Schadenfreude*, or the joy one experiences at the failure of another, often is directed toward previously successful others not perceived as deserving of success. These others may have attained their wealth or status through illegal means, or luck, or inheritance. In contrast to this, the positive emotion of gratitude is experienced when someone intentionally or volitionally helped another attain a goal and succeed. Unintentional or forced help will not give rise to gratitude. Even jealousy is not likely to be aroused if the behavior of the desired other was not subject to volitional control—one is jealous when a mate stopped the elevator in mid-flight with a rival but not when the elevator had a breakdown at that point. That is the logic behind the emotional classification in Table 2.8.

Personality Inferences and the Moral Emotions

While I indicated a desire to examine emotional reactions to the success of others, only two such affects are evident in Table 2.8: envy and admiration, which I reasoned to be respectively linked to ability and effort. I also then began to consider what individuals actually communicate when they succeed because of ability or effort. It is know from the vast amount of research on excuses that individuals alter their public accounts for failure so they are held not responsible. Are similar communication strategies at work given success? And what are likely to be the inferences if an individual states he or she succeeded because of high ability or because of effort expenditure?

Table 2.9 shows the hypothesized emotional reactions and personality inferences if ability and effort are actual and communicated causes. Table 2.9 reveals what might be considered a paradox: we envy ability, yet if another states he or she has it we react negatively with an accusation of arrogance. In contrast, admiration and modesty form a consistent pairing given an effort attribution (see Hareli & Weiner, 2000, 2002a, 2002b).

Table 2.9.   Attribution-Related Affects and Personality

Cause	Success Outcome	
	High Ability	*High Effort*
Affect reaction	Envy	Admiration
Personality inference if communicated	Arrogance	Modesty

I now have completed a number of studies on these emotions, personality inferences, and their attributional underpinnings and moral connections (Hareli, Weiner, & Yee, 2004). The findings tend to be complex (perhaps because of the experimental designs) and I find them quite intriguing. In the exemplar study I have chosen to present, we told our participants someone had succeeded because of one of four reasons: high ability, high effort, good luck, or help from friends. We then crossed this information with what the person communicated to peers when asked the reason for success, using the same four causes. Thus, for example, the person is reported to succeed because of high ability and says that; or succeeds because of high effort but communicates the success was due to high ability, and so on. We then obtained ratings of the emotions of envy and admiration as well as personality inferences of arrogance and modesty. We also asked our participants how honest the achiever was. This resulted in a host of additional speculations about the relation between honesty and the reactions resulting from causal beliefs. For example, it is known that statements of personal ability are regarded as arrogant, but is this also the case when ability is the real cause? Individuals prefer honest communications, so perhaps a true claim of high ability will not be regarded as arrogant. In a similar manner, perhaps a true claim that success was due to effort or an external cause such as good luck will not be regarded as modest. But one can make the reverse argument as well, which is that a statement regarding one's high ability as the cause of success elicits beliefs of arrogance regardless of its truth value. Similarly, one may be regarded modest when reporting an external cause of success even when that is the true cause.

Figure 2.4 depicts the effects of the actual cause on feelings of envy. The actual cause was the most dominant determinant of this emotion. A

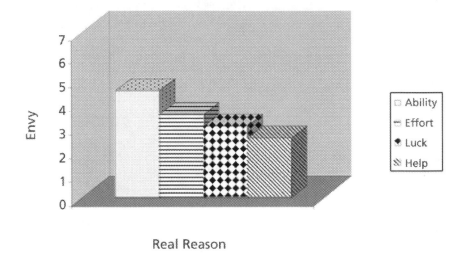

Figure 2.4.   Reported envy as a function of the real cause of success (from Hareli, Weiner, & Yee, 2004).

suggested already, envy is augmented given internal causes for success and is greatest given ability as the cause. Nonetheless, it should be noted that successful effort also arouses envy (but perhaps only because of the success), which I had not fully anticipated.

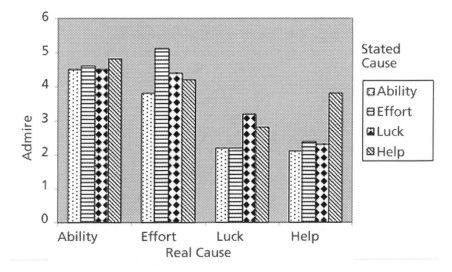

Figure 2.5.   Admiration as a function of the real and communicated (stated) causes of success (from Hareli, Weiner, & Yee, 2004)

The results regarding admiration are more complex and less supportive of my original belief (see Figure 2.5). Admiration is again primarily
determined by the actual cause and is greater given real internal than
external causes of success. However, unlike envy, one is admired as much
for succeeding due to effort as because of ability. Admiration is thus
linked to both controllable and uncontrollable internal causes of success,
suggesting it may be less of a moral emotion than I anticipated. However,
the most admired individual succeeded because of high effort and stated
this was the case, so perhaps there is an ability/effort distinction in
regards to admiration.

Perhaps more intriguing are findings in regard to judgments of arrogance and modesty and their relation to honesty. Figure 2.6 depicts the
data regarding arrogance. As expected, communicating the two internal
causes of ability and effort is regarded more arrogant than communicating
external causes, and ability communications are considered more arrogant
than statements of effort expenditure. These findings closely mirror the
results for envy. You can also see in the graph that real causes play some
role in arrogance inferences, for when ability is the real cause (the left bar
graph in all four groupings), inferences of arrogance are reduced. However, when ability is the stated cause, then inferences of arrogance are high
and do not significantly differ as a function of the real cause. Thus, contrary to some beliefs, honesty does not mitigate perceptions of arrogance
when ability is stated as the real cause. In other research we have found

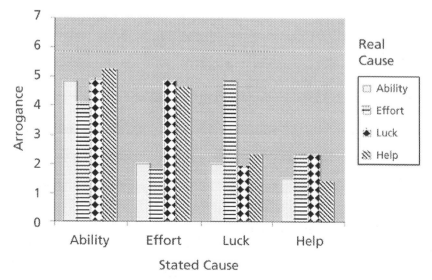

Figure 2.6.   Arrogance as a function of the real and communicated (stated) causes
of success (from Hareli, Weiner, & Yee, 2004).

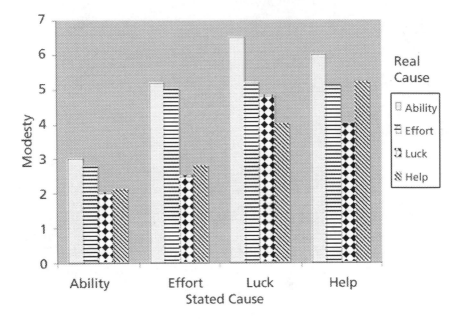

Figure 2.7.   Modesty as a function of the real and communicated (stated) causes of success (from Hareli, Weiner, & Yee, 2004).

that even if success is exceedingly difficult and atypical, including the winning of a Nobel award, stating high ability is regarded arrogant. Being Einstein does not allow you to say "I am an Einstein."

Turning to effort, here we find that stating effort when it is the real cause is not regarded arrogant. But given effort as the stated cause, arrogance is augmented when the real causes are external to the person (luck and help). Considering for a moment external stated causes of success, inferences of arrogance are low except when luck is claimed, although effort is the real cause. Perhaps in this instance luck is interpreted a characteristic of the person and thus akin to ability in being an internal and stable cause.

What about modesty? The modesty data are shown in Figure 2.7. The data indicate that stating external causes increases modesty. In addition, if the real cause is internal, and particularly ability (the left bar graphs in each grouping), then modesty is augmented. Real causes play a greater role in judgments of modesty than of arrogance. If one has little ability, then there is a reduced opportunity to be perceived as modest. On the other hand, one always can be regarded as arrogant, regardless of the truth. Like arrogance, however, honesty regarding ability does not greatly mitigate impressions of lack of modesty.

Finally, let me point out some interesting data about honesty, which are shown in Figure 2.8. As expected, in all conditions when the stated cause matches the real cause, then the communicator is perceived as most honest. However, not all lies are the same. The combination of communicating ability or effort with the real causes external to the achiever is particularly dishonest. This is more dishonest than, for example, reporting external causes when the real cause of success is internal to the achiever.

In sum, honesty does not mitigate inferences of arrogance in the case of ability self-statements, and even being least honest (communicating ability when external causes are real) does not increase inferences of arrogance. Similarly, impressions of modesty are not reduced merely because the achiever is telling the truth. Finally, inferences of honesty are not divorced from beliefs about arrogance and modesty—an arrogant liar is more dishonest than a modest liar, probably because one questions the motives that gave rise to the lie.

A CONCLUDING NOTE

In this chapter I have presented my most recent thoughts and research regarding an attributional approach to social motivation, justice, moral

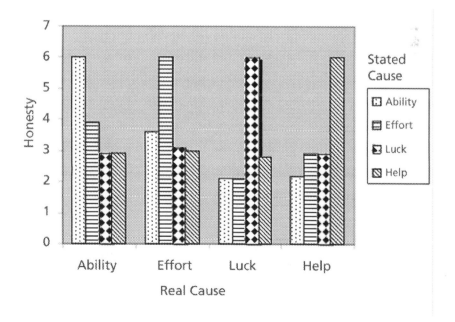

Figure 2.8.   Honesty as a function of the real and communicated (stated) causes of success (from Hareli, Weiner, & Yee, 2004).

emotions, and personality inferences. This approach focuses on perceived causes of outcomes, relations of causal perceptions to beliefs about the person, associations between perceived personal inferences and emotions, and then links between thinking, feeling, and acting. The specific topics examined included achievement evaluation, stigmatization, help-giving, and aggression; a variety of affects including admiration, anger, envy, scorn, and sympathy; impression management techniques; and the personality inferences of arrogance, modesty, and honesty. The theory is indeed rich and certainly much more is included than has been presented here.

These also are very fertile topics for organizational psychology. Any in the list mentioned above has its place within an organizational framework. Under what conditions is help given? When is aggression expressed and why? How do others feel about successful colleagues? How do these colleagues present themselves? Pursuing these and related topics can go on for a long time, which is what I believe will happen. Attribution theory is not about to go away, given the relational fertility spawned by the constructs. So, I suspect many more books with this topic lie in the future.

REFERENCES

Hareli, S., & Weiner, B. (2000). Accounts for success as determinants of perceived arrogance and modesty. *Motivation and Emotion, 24,* 215–236.

Hareli, S., & Weiner, B. (2002a). Dislike and envy as antecedents of pleasure at another's misfortune. *Motivation and Emotion, 26,* 257–277.

Hareli, S., & Weiner, B. (2002b). Social emotions and personality inferences: Ascaffold for a new direction in the study of achievement motivation. *Educational Psychologist, 37,* 183–193.

Hareli, S., Weiner, B., & Yee, J. (2004). *Do truthful accounts mitigate perceptions of arrogance and modesty?* Unpublished manuscript, University of California, LosAngeles.

Heider, F. (1958). *The psychology of interpersonal relations.* New York: Wiley.

Ickes, W. (1996). On the deep structure of attribution-affect-behavior sequences. *Psychological Inquiry, 7,* 236–240.

Izard, C. E. (1977). *Human emotions.* New York: Plenum Press.

Kelley, H. H. (1967). Attribution theory in social psychology. In D. Levine (Ed.), *Nebraska Symposium on Motivation* (Vol. 15, pp. 192–238). Lincoln: University of Nebraska Press.

Rudolph, U., Roesch, S. C., Greitemeyer, T., & Weiner, B. (in press). A meta-analytic review of help giving and aggression from an attributional perspective: Contributions to a general theory of motivation. *Cognition and Emotion.*

Weiner, B. (1991). Metaphors in motivation and attribution. *American Psychologist, 46,* 921–930.

Weiner, B. (1995). *Judgments of responsibility: A foundation for a theory of socialconduct.* New York: Guilford Press.

CHAPTER 3

ATTRIBUTIONS AND THE ACTION CYCLE OF WORK

Terence R. Mitchell, Thomas W. Lee, Dong-Yeol Lee, and Wendy Harman
University of Washington Business School

ABSTRACT

We present a theory that integrates a number of divergent motivational and attributional aspects of goals. It focuses on how assigned deadline goals influence the behaviors of pacing (resource allocation to a particular task) and spacing (resources allocated across tasks). These activities occur while one is actually striving to reach multiple deadline goals. The components of the theory include plans, deadlines, feedback, goal discrepancies, and pacing and spacing. The key variable for our attributional analysis is goal discrepancies. We argue that when one is significantly ahead (a positive goal discrepancy) or behind (a negative goal discrepancy) that attributions and emotions are the psychological reactions that cause subsequent action. We present propositions about how this process unfolds. We close with a discussion of how attributions are important for understanding the dynamic nature of work.

Oftentimes, people strive to achieve or fail to achieve their work goals. In some cases, these successes or failures are important and significant. People typically try to make sense of these outcomes, and they frequently employ attributional processes to do so. This chapter describes the moti-

Attribution Theory in the Organizational Sciences, pages 25–48

vational processes that may occur when individuals work toward goal attainment. While many of our ideas and propositions come from existing theory, our focus is on three topics infrequently discussed or tested. First, we are interested in assigned deadline goals. These goals are externally imposed and contain a time criterion for success. Second, we focus on a multiple goal context; people commonly work on several tasks at once. Third, our criteria are behaviors that involve goal striving and not goal accomplishment. More specifically, our focus is on the resources (time and effort) that people allocate to a single task (which we call pacing) and the resources they allocate across tasks (which we call spacing). Thus, the theory examines how assigned deadline goals for multiple tasks influence attributions, resource allocations, and behaviors involved in striving to reach these goals.

We recognize that this focus is narrow. It is sufficient to say at this point that no empirical studies have been done on this topic, and as we shall see, the topic is very complex, even with these limiting parameters. More important, this theory is relevant for millions of people at work. Secretaries are a good example. They have multiple tasks with assigned deadlines. They work independently and have considerable control over their resource allocations, making volitional pacing and spacing decisions throughout the day. In short, it is a neglected but important topic.

The chapter is divided into two main parts. The first part presents the overall motivational model. It provides the foundation for the second part, where attributions play a key role. The second part describes how attributions moderate the relationships between positive or negative goal discrepancies (we are way ahead or behind schedule to reach a deadline goal), and the amount of time and effort we allocate to reach these goals.

This latter section is the "heart" of the model. The early work on attribution theory (Heider, 1958; Weiner, 1977) has emphasized how people use attributions as a causal sense-making mechanism. This process is especially important for events that are unexpected, novel, and usually (although not always) negative (Weiner, 1990; Weiner & Graham, 1999). We elaborate on the different ways people can interpret performance discrepancies, their emotional reactions, and subsequent decisions to continue to try to reach a goal or to give up on the goal and switch to another activity. It is this dynamic attributional process that we believe will help to explain the action cycle of work: What people actually do when striving to reach their goals.

PART 1: THE MOTIVATIONAL MODEL

Theoretical Foundations

The research involving goal constructs is enormous (Austin & Vancouver, 1996) and includes such areas as motivation, learning, personality, and performance. Recently, research on goals has been divided into two general camps: those studies that focus on the antecedents to action and those that focus on action itself—a goal setting versus a goal striving distinction (Kanfer & Kanfer, 1991; Lord & Levy, 1994). In the area of organizational studies, the paradigm that has dominated the former approach is Locke and Latham's (1990) goal-setting theory. Most of this research focuses on how goals are chosen and the content of the goal.

Since our interest involves only one goal type (short-term, lower level, specific task goals that are assigned and time related), much of this work does not apply to our analysis. The relevant points, however, are as follows. First, motivational processes are seen as initiated by goals. Locke (1994) sees goals and the aspirations to reach them as the basic force for life itself. Second, when there is a discrepancy between the goal and our progress, we are motivated to close that gap (Locke & Latham, 1990). These comparisons result in emotional reactions as well as cognitive evaluations of competence and attributions (Latham & Locke, 1991). Third, feedback (which provides discrepancy information) is necessary for goals to have a motivational impact (Erez, 1977). Fourth, goals that are difficult to reach (once accepted) and are clear and specific are more motivational than vague or easy goals (Locke & Latham, 1990). Fifth, commitment to a goal is necessary and implies that more resources will be allocated to a goal to which one is committed (Diefendorff & Lord, 2000). Finally, goal setting involves antecedents to action, such as the plans or strategies constructed to reach the goal (Locke & Latham, 1990), as well as the consequences of attaining or failing to reach a goal, such as subsequent goal revisions (Klein & Dineen, 2002). We focus on what happens between goal setting and goal attainment, which is called goal striving.

Three theoretical approaches present ideas that help inform our analysis. First, the work of Kanfer and her colleagues (Kanfer, 1996; Kanfer & Ackerman, 1989; Kanfer & Heggestad, 1997) focuses on resource allocation and skill acquisition; how people learn new skills over time. Their theory is about how attentional resources are allocated over the learning curve and how individual differences in ability and personality can influence this process. They draw heavily on the principles of self-regulation where people are seen as actively controlling their cognitive and emotional psychological processes for the purpose of attaining goals (Diefendorff & Lord, 2000; Gollwitzer & Bargh, 1996). This allocation process is central to our analysis. Second, control theory (Campion & Lord, 1982; Klein, 1989; Lord & Hanges, 1987; Lord & Levy, 1994) emerged from work on control systems and cybernetics (Carver & Scheier, 1981; Powers, 1973) and focuses on goal discrepancies. The Test, Operate, Test, and Exit (TOTE) unit described by Miller, Galanter, and Pribraum (1960) explains how behavior may change as one approaches a goal (Markman & Brendl, 2000). Goal striving is a dynamic process that occurs over time (Klein, 1989). Research by Lord and colleagues extends these ideas to investigate multiple tasks (Kernan & Lord, 1990) and the importance of planning as a self-regulatory process that facilitates the successful response to discrepancies and multiple goal attainment (Diefendorff & Lord, 2000). Finally, action theory (Frese & Zapf, 1994; Gollwitzer, 1996; Kuhl, 1984) emphasizes goal striving as part of an "action cycle" involving a preaction phase including goals and plans and an action phase where one pursues the goal. This latter phase includes momentary and ongoing changes in thought and action and ways one can control this process. Kuhl (1984, 1986), for example, emphasizes that the thoughts, attributions, and emotions involved in the planning process may differ significantly from those that occur while working and that one of the major functions of planning can be to deflect us from disruptive thoughts and negative emotions that occur during goal pursuit. We focus on the changes that occur while working toward a goal.

These approaches to goal striving share some elements. They all see the process of goal attainment as cyclical and dynamic. There are multiple phases, including: (1) goal choice; (2) plans and strategies; (3) actual work on a task involving progress checks and responses to discrepancies such as attributions; and (4) a final task performance stage. We focus on the planning and action stages, the first of which involves the anticipated allocation of resources (time, effort) while the second involves the actual expenditure of these resources. Also critical is the idea that the psychological and motivational processes governing one's plans and intentions are different from those that govern our action. Our theory attempts to integrate these ideas with a focus on assigned deadline goals, multiple tasks, attributions, and the behaviors of spacing and pacing.

Pacing and Spacing: Component Parts

Pacing and spacing decisions, as a result of attributions, are the behaviors of interest or the outcomes of our analysis. They obviously mediate the relationship between goals and performance, but for our analysis, they are the dependent variables. Before proceeding with predictions and propositions, however, we need to define the constructs included in this theory. As an initial overview, we describe the work of a secretary. When the day begins, he already knows the existing tasks that await him; he also knows the deadlines that accompany these tasks; and he forms a rough plan about how to proceed in terms of allocation of resources (time, effort) within and across tasks. As the day unfolds, some tasks are completed and new ones arrive (some expected and some unexpected). He is also aware, on occasions throughout the day, of his progress on one or more tasks with respect to the deadlines. These assessments, including attributions, may cause him to speed up or slow down on a particular task, switch tasks, or revisit the plan and reallocate his resources. These are pacing and spacing behaviors. The major components of this process are the deadline goals, the plan, the progress or goal discrepancy judgments, work context variables like the number of tasks to be done, and the pacing and spacing decisions.

Pacing. Pacing is the allocation of resources to a single goal, and it involves both time and effort (Kernan & Lord, 1990). At an initial planning phase, it involves the anticipated time to be spent on a task. At a later action stage, it is the amount of time and effort exerted over time (rate), thus pacing can involve both time and effort. In response to a goal discrepancy or some other event (e.g., an interruption), depending on the attribution, one can change the speed at which he or she is working (the rate of effort); one can change his or her allocation of time (add or subtract); or change both. We would add that thinking about and developing task strategies takes time and is therefore part of the pacing activity. Thus, pacing during the action stage refers to the rate of effort as well as to changes of its component parts of time and effort.

Spacing. Spacing involves the allocation of resources across goals. During the planning stage, spacing again involves time allocations. It is the distribution of total available time to the tasks that await completion. During the action stage, spacing involves the decision to redistribute time across tasks and/or change from one task to another. This latter action is called switching; ceasing work toward one goal and working toward another goal. While spacing is an important action in the real world and merits inclusion in any overall theory of work behavior, we concentrate primarily on how attributions are involved with decisions about pacing.

Assigned Deadline Goals

The criterion for goal attainment for assigned deadline goals is task completion by a particular time. Four goal attributes that cause pacing and spacing are reviewed by Austin and Vancouver (1999): (1) importance–commitment; (2) difficulty; (3) specificity; and (4) temporal range.

To these four, we will add two new variables (urgency and accountability) that are relevant for attributional and motivational responses to deadlines.

Importance–Commitment. This construct involves two major components: the determination to try for the goal and one's persistence in trying to reach the goal when confronted by obstacles (Diefendorff & Lord, 2000; Hollenbeck & Klein, 1987). The research on importance–commitment focuses on the expected value of goal attainment (Kernan & Lord, 1990). People think about the personal consequences of reaching a goal, and the overall and relative (compared to other competing goals) goodness or badness of these consequences influences their commitment to the goal (Karniol & Ross, 1996; Latham & Locke, 1991).

Importance accrues from factors other than personal outcomes. For example, priorities often come from the workplace itself. People may also prefer some tasks or people (for whom they do work). Thus, some dimensions of importance–commitment (e.g., liking the task) may be independent of other dimensions (e.g., work priority). The data suggest that higher importance–commitment leads to higher motivation and the allocation of more resources (Latham & Locke, 1991). In addition, importance–commitment partly determines how one reacts to a goal discrepancy, the attributions that occur, and the decisions made (Cropanzano, James, & Citera, 1993).

Difficulty. The more difficult the goal (with sustained commitment) the greater the motivation and allocation of resources to a goal. This applies to deadlines as well as other types of goals (Latham & Locke, 1991). Note, for deadlines, time is part of the definition of goal difficulty. That is, when a new task is assigned with a deadline, a worker can assess their chance of reaching such a goal. This estimate will reflect the time and resources that *can be* allocated to this task, *given all the other tasks to be done.*

Goal Specificity. Deadlines can be vague (sometime next week) or specific (tomorrow at 10:00 AM). Specific goals are more motivating than vague goals (Locke & Latham, 1990), and people will work harder on tasks with specific goals rather than tasks with vague goals (Austin & Bobko, 1985). This should be true for assigned deadlines.

Temporal Range. Goals that are proximal (with task importance held constant) are more motivating than goals that are distal (Kanfer & Ackerman, 1989; Kanfer & Kanfer, 1990). Since deadlines by definition include differences in time, these results should hold for deadline goals (Sears & Woodruff, 1997). Besides the motivational forces described above that result from goal importance, difficulty, specificity, and temporality, there are two additional states that result from deadlines: a sense of urgency and felt accountability (a dimension that is inherent in attributions).

Urgency. Deadlines create an additional sense of urgency due to their criterion of completing a task by a certain time. People report that deadlines press on them, that tasks with deadlines insist on action, that they seem more important or urgent (Covey, 1989). "Increased urgency—in the face of an approaching deadline—may represent an additional requirement for commitment to future action" (Heckhausen & Kuhl, 1985, p. 136). Thus, urgency can influence commitment to a goal, one's intentions to work toward that goal, as well as their actions in trying to reach the goal.

Accountability. We are focusing on *assigned* deadline goals. Because the goals are assigned and because deadlines have a relatively easy-to-observe criterion based on time, we believe they invoke feelings associated with the idea of accountability. Accountability is defined as "calling to give accounts to another for deviation between the event for which one is responsible, and relevant expectations" (Cummings & Anton, 1990, p. 628) and is a key attributional dimension as well (Weiner, 1990). Frink and Klimoski (1998) say that two key elements to induce accountability are: (1) a clear standard (the deadline) and (2) being held to account by someone else (the person who assigns the deadline). These elements produce a felt responsibility to reach the goal (Cummings & Anton, 1990) and a concern over being evaluated by others, frequently labeled as evaluation apprehension (Frink & Klimoski, 1998). Thus, assigned deadlines convey additional motivational forces over and above nondeadline goals.

Plans

Once one has an overview of their tasks and deadline goals, they form a plan. Each task or goal is associated with its context, time, and strategy (tactics) to form an intention (Ajzen & Fishbein, 1980; Tubbs & Ekeberg, 1991), and these intentions are combined to form an overall plan (Earley, Shalley, & Northcraft, 1992). Because most tasks, in the types of jobs we are focusing on, are short term, each day some things are likely to be completed and new tasks are likely to be assigned. Thus, new plans are probably made at the start of every day (Klein, 1989).

The plan represents a cognitive representation of future states and situations. It reflects our priorities (Covey, 1989) and contains details, allocations, and time frames (Frese & Zapf, 1994). Thus, plans are the bridge between our deadline goals and our actions of goal striving. The plan tells us how to proceed, what to do first, what to do next, and how much time to allocate to each task (Earley, Wojnaroski, & Prest, 1987). Both the goals and the plans also influence the action phase. They help us get started (Gollwitzer & Schaal, 1998), provide shields against distractions (Bargh & Gollwitzer, 1994; Lord & Levy, 1994), and help us persist in the face of a goal discrepancy (Earley et al., 1987). From our perspective, the plan is related to spacing and pacing in two ways. First, the plan involves the initial allocation of resources to a particular task as well as across tasks. Second, once one begins working on their plan, things can happen that cause one to change their original intentions about these allocations. They may speed up or slow down (pacing), change time allocation to one task (pacing) or multiple tasks (spacing), or change from one task to another (switching). The main mechanism that causes these changes in plans is the goal discrepancy.

Goal Discrepancy

At various times during the day one can assess their progress toward a particular goal. For deadlines "individuals are likely to notice how much of their allotted time has elapsed before the deadline arrives" (Waller, Conte, Gibson, & Carpenter, 2001, p. 588). The comparison here is with the allotted time (the plan), not the absolute amount of time, and this comparison functions as feedback (Kluger & DeNisi, 1996). The discrepancy (either positive or negative), if large enough, results in different types of reactions. It focuses attention (Lord & Hanges, 1987), prompts attributional processes (Thomas & Mathiew, 1994) and generates feelings of satisfaction or dissatisfaction (Bandura & Cervone, 1986) and positive and negative emotions (Karniol & Ross, 1996; Martin, Tesser, & McIntosh, 1993). What is critical theoretically is that the discrepancy invokes a thoughtful mode that focuses attention on a particular task and on how one is performing (Cropanzano et al., 1993). This attention can result in changes in motivation (Kuhl, 1996) and increases the salience of a particular goal (Carver & Scheier, 1981). In addition, if the discrepancy is too large, it may lead to a disengagement from that task (e.g., switching) and starting work on a different task (Kernan & Lord, 1990).

Work Context

The final components for this theory focus on attributes of the work setting. These are contextual variables that influence the psychological processes and behaviors involved in pacing and spacing. The number of tasks can influence both the initial allocations in the plan as well as pacing and spacing later on. In addition, the number of "new tasks" brought to the person during the day may cause changes in the allocations. And finally, interruptions or unanticipated events can result in reallocations and pacing and spacing behavior (Kanfer & Kanfer, 1991). Waller and colleagues (2002), in their analysis of groups, point out that "a sudden interruption, such as a deadline change, creates a pause in activity and the opportunity for the members of a group to stop and think about its task progress" (p. 1047).

PART 2: PROPOSITIONS AND ATTRIBUTIONS

This next section focuses on how attributions enter into both the planning and action stages of the work cycle. Attributions are used as psychological processes to help an individual assess the causes of their own behavior and the behavior of others (Heider, 1958). Since our work focuses on the individual, working independently, we can concentrate on self attributions. Thus, much of Kelly's (1967) work focusing on the dimensions of the distinctiveness, consensus, and consistency of another person's behavior is omitted while the work of Weiner and his colleagues (Weiner, 1985, 1990; Weiner & Graham, 1999) is central to our analysis. We describe how attributions based on the dimension of locus of causality, stability, and controllability influence reactions to goal discrepancies.

Initial Time Allocations (Anticipatory Pacing and Spacing): The Plan

What confronts our hypothetical worker initially is multiple tasks with multiple assigned deadlines and the need to form a plan to deal with these tasks. This plan involves an estimate of *task importance* for each task (and the things that comprise this judgment like personal outcomes, organizational priority, or intrinsic interest in the task), and the *difficulty* and *specificity* levels of the deadline goal for each of the tasks to be completed. These deadlines also create the two psychological states we discussed, namely, urgency and felt accountability. These factors are combined into a judgment for each task, which we label the overall *commitment to meet the deadline*. The person then looks at this array of tasks and deadlines, as well as their commitment to the deadlines and the time available to reach them, and makes allocation judgments for the coming day (or parts of the day) and vaguer, less precise judgments for following days. Figure 3.1 represents this process.

Some obvious predictions are based on this process. First, tasks with assigned deadlines are allocated more resources than those without deadlines. Second, proximal, important, and specific deadlines are allocated more resources than distal, unimportant, and vague ones. Third, the disposition of time urgency will moderate these relationships, and fourth, monitoring will also moderate these relationships (by increasing one's sense of responsibility).

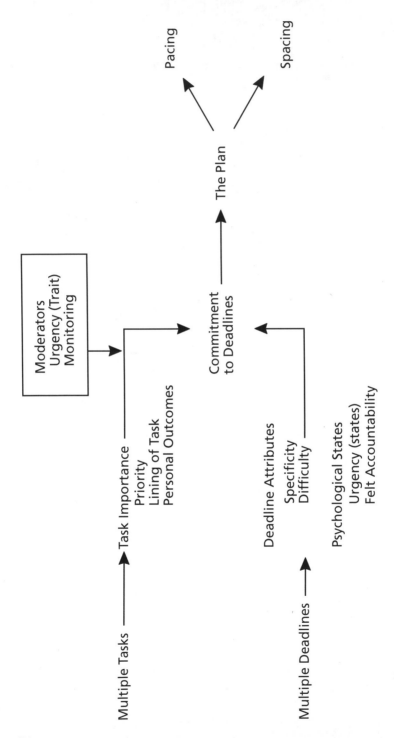

Figure 3.1.   Pacing and spacing: Initial allocation.

These propositions focus on the *intended* allocation of resources. Although straightforward, they have not been tested with initial allocations as the dependent variables. These allocations represent what traditional theory typically use as the independent variables in most goal-setting studies—what people say is their goal and intention. We believe that what people encounter while they are striving for goals can change their intentions and that these new intentions are better predictors of their actual actions than their initial intentions. Attributions play a key role in this process.

Time and Effort Expended during the Action Cycle (Pacing)

Once work commences, the plan should serve as an initial guide and predictor of actual effort expended; that is, specific, proximal deadlines will lead to more actual effort exerted and time spent on a task than vague, distal deadlines. The key variable as time passes is one's judgments about whether they are actually going to meet the deadlines as anticipated (Waller et al., 2001). This is the goal discrepancy judgment. Because one has multiple tasks and deadlines, the goal discrepancy or progress judgment is fairly complex. It is not simply how much time is left combined with how much work has already been completed on a task. It is that information compared to how much time was allocated in the plan. This latter judgment is based on the allocations to all the other tasks, and the complexity comes from having multiple deadline goals.

Most research on goal setting, because there are no multiple tasks, simply looks at the discrepancy between the deadline and current progress on a task. We are suggesting that the person assesses if they are on track, given everything else that needs to be done. This is an estimate of their probability of successfully meeting the deadline and results in three broad categories of inference: I'm on track, I'm behind, or I'm ahead. Naylor and Ilgen (1984) call it a feasibility evaluation. Pacing, as a change in the allocation of resources to a particular task, occurs in response to this judgment and attributions play a key role. Figure 3.2 represents these processes.

On-Track. An on-track judgment suggests that the initial allocation was appropriate as one gauges their progress on a particular task. However, some authors suggest that as we get close to the goal we increase our effort (Karniol & Ross, 1996; Smith & Lem, 1955; Waller et al., 2001). "Work indicates that individuals are likely to increase task activity before the deadline arrives" (Waller et al., 2001, p. 588). Thus, when one is "on track" to meet a deadline they will speed up (rate of effort) as they get close to the deadline. This process does not appear to be mediated or moderated by attributions.

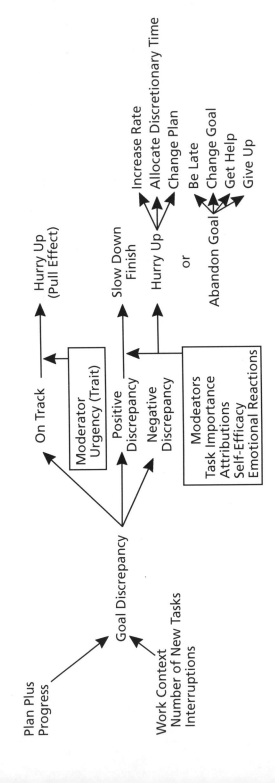

Starts Process   Status   Behavior

Plan Plus
Progress

Goal Discrepancy

Work Context
Number of New Tasks
Interruptions

On Track → Hurry Up
(Pull Effect)

Moderator
Urgency (Trait)

Positive
Discrepancy → Slow Down
Finish

Negative
Discrepancy → Hurry Up

Modeators
Task Importance
Attributions
Self-Efficacy
Emotional Reactions

or

Abandon Goal

Increase Rate
Allocate Discretionary Time
Change Plan

Be Late
Change Goal
Get Help
Give Up

Figure 3.2. Pacing: Changes in Allocation

36

Negative Goal Discrepancy. Frequently, our estimate is that we are behind in our progress to reach the deadline, given our planned allocation of resources (Donovan & Williams, 2003). This is called a negative goal discrepancy (NGD). NGDs occur frequently due to a general tendency to underestimate how long it will take us to accomplish a task (Buehler, Griffin, & Ross, 1994). Note that NGDs are unexpected and negative and are likely to prompt a thoughtful causal analysis.

Initially, a NGD draws our attention to that task (Cropanzano et al., 1993) and our reactions are to try to reach, or to give up on, the deadline (Blount & Janicek, 2001; Campion & Lord, 1982). People differ in their reactions to a NGD and such differences are moderators of the NGD-allocation relationship (discussed below). More specifically, cognitive assessments are important, such as the importance of the task, one's probability of reaching the deadline, and attributions about why one is behind. In addition, there are emotional reactions to a NGD.

With a given NGD, the decision about whether to allocate more or less resources to meet the deadline depends on the importance of the task. For example, Emmons and Diener (1986), Donovan and Swander (2001), and Cropanzano and colleagues (1993) suggest that when the task is important, a person is more likely to allocate more time and effort than to give up. Hollenbeck and Williams (1987) empirically demonstrated that more effort was exerted in the face of a NGD under conditions of high goal importance than low goal importance.

Another important moderator is our perceived efficacy for successfully meeting the deadline (Kuhl, 1984). If this estimate is over some threshold, people should increase their resource allocation and still try to reach the goal (Bandura & Cervone, 1986; Donovan & Swander, 2001; Kanfer & Ackerman, 1989). This increase can occur in a number of ways. We can speed up (e.g., increase our rate), allocate more time but leave the initial plan unchanged for other tasks (e.g., have a shorter lunch hour, stay late), or revisit the plan and redistribute our time allocation to this task. However, when efficacy is judged to be low, the person may abandon the deadline goal (Bandura, 1997). In this case, they can simply decide to be late (and take the consequences), try to change the goal by talking to the person who set the deadline, get help (get someone else to meet the deadline), or simply abandon the task and move on to some other task (Klein, 1989). Kluger and DeNisi (1996) and Cropanzano and colleagues (1993) suggest that repeated or large negative feedback discrepancies lead to goal abandonment.

In response to a NGD, the individual may also make an attributional judgment (Weiner, 1985); that is, they search for the reasons why they are behind (Klein, 1989). These reasons are usually classified along three dimensions: locus, stability, and control. Most researchers agree that attribution judgments moderate the reaction to the NGD (Locke & Kristof, 1996; Thomas & Mathiew, 1994; Williams, Donovan, & Dodge, 2000) and influence whether one will try for the goal or give up the goal.

Proposition 1:   *Attributions moderate the NGD-allocation relationship.*

A more specific proposition utilizes the locus dimension. Since effort is likely to make a major contribution to performance on these types of tasks, if the individual believes he or she is responsible for being behind (an internal attribution), he or she will allocate more resources to reaching the goal than if an external attribution is made (Cropanzano et al., 1993; Klein, 1989). For example, a long lunch break may result in an internal attribution for a NGD (I did not put in enough time on this task) while a computer malfunction may produce an external attribution. Staw suggests that the "individual must feel personally responsible for the negative consequences" (1976, p. 28) to increase their effort.

Proposition 2:   *Locus of causality moderates the NGD-allocation relationship. An internal attribution to lack of effort will lead to an attempt to reach the goal while an external attribution is likely to lead one to give up on the deadline goal.*

The controllability dimension is also important. Carver and Scheier (1981), Ford (1982), and Williams and colleagues (2000) suggest that, based on Seligman's (1993) work, attributions to an uncontrollable cause of failure will lead to lower expectations of future success and reduced commitment to reaching a goal. Note that this attributional effect appears to occur through an expectation of success or efficacy type judgment.

Proposition 3:   *Perceived controllability moderates the NGD-allocation relationship. An attribution to uncontrollable causes of the NGD is likely to lead to giving up while perceived controllable causes will lead to renewed attempts to reach the goal.*

The stability dimension may also be relevant. Thomas and Mathiew (1994), for example, suggest that an attribution to an *unstable* cause for a NGD (e.g., bad luck) will not influence one's self-efficacy very much and therefore leave resources allocated unchanged. In addition, Williams and colleagues (2000) suggest that a *stable* attribution for a NGD (e.g., a tough task) may *decrease* one's expectations of future successes and therefore lead to a decrease in resources. In combination these ideas would suggest the following.

Proposition 4:   *Perceived stability moderates the NGD-allocation relationship. An attribution to a stable cause for a NGD will lead to giving up on the deadline while attributions to unstable causes are likely to produce increased resources to reach the goal.*

There are also some emotional reactions to NGDs. "Non-attainment in the expected time should be experienced as undesirable, giving rise to negative effect" (Kruglanski, 1996, p. 610). Some authors see these reactions as a result of the cognitive activity described above (probability of success, task importance, efficacy, and attributions) while others suggest there is a direct NGD-emotional reaction (Bargh & Chartrand, 1999). Klinger (1996), for example, argues that emotions happen quickly as a reaction to a physical or psychological event (within 300 milliseconds) and that the cognitions–emotions causal arrow goes both ways. In this case, all NGD's would initially result in negative affect. Regardless of the causal direction, emotional reactions to a NGD or the attribution about it are usually negative, including guilt, shame, anxiety, anger, negative mood, and depression (Carver & Scheier, 1981; Cropanzano et al., 1996; Ford, 1992). Guilt, shame, and anxiety are suggested to be more likely to lead to positive allocations and trying harder while anger, negative mood, or depression may lead to lower allocations (Cropanzano et al., 1993; Ford, 1992). In other words, different emotions are associated with different self-regulatory processes (Higgins, 1998).

Finally, the work context can also influence the amount of pacing by making it more difficult to follow the time estimates in the initial plan allocation. Receiving new tasks and experiencing interruptions should increase the goal discrepancies, reassessment, and reallocation that occur during pacing (Kanfer & Kanfer, 1991). They are also likely to result in external and uncontrollable attributions.

Proposition 5:   *An external attribution will occur for an interruption causing one to reduce their efforts to reach the deadline goal.*

One last point about NGDs should be mentioned. As indicated above, the responses to a NGD are complex and may involve multiple psychological mechanisms. However, we believe that these processes are probably associated with one another. If someone is behind because they took too long a lunch, they are likely to make an internal attribution, feel some guilt or responsibility, retain their self-efficacy, and probably increase their efforts (e.g., their rate of effort) or time (e.g., stay late) to complete the task. If, however, the NGD is due to a computer failure or interruption, they may make an external attribution, feel angry or discouraged, and decrease their effort or move on to another task. In short, many of these moderators may be interrelated.

Positive Goal Discrepancy. Sometimes, we are way ahead of schedule, and we determine we have allocated more resources than we need to finish a task. This is a positive goal discrepancy (PGD). We could finish and allocate the extra time to another task (likely if this task is very important), but most authors suggest that we will lower our sights (Campion & Lord, 1982), for example, let the time expand to meet the allocation (Lim & Murnighan, 1994) or adjust our behavior to fit the time (McGrath & Rotchford, 1983). Waller and colleagues (2002) show that when groups perceive "a deadline change as meaning an increase in time resources, creating a situation of 'time abundance,' the group may be less motivated to increase task performance activity" (p. 1048).

Our analysis suggests that attributions will again be important and may make our predictions more complex than those suggested above (i.e., people slow down with a positive goal discrepancy). Two internal attributions are possible: "I'm ahead because of my effort and skill," or "I'm ahead because I misjudged the difficulty involved in reaching the deadline." We believe the former internal attribution will lead individuals to finish early and switch to another task, while the latter will lead individuals to slow down.

The first of these ideas is based on the premise that finishing early due to your own efforts or abilities is a positive consequence. It can function as a self-congratulatory response (good work, you finished early) and can also increase one's self-efficacy for a particular task. The only potential drawback to finishing early is that if one's actual completion times are being monitored, more work may be assigned to someone who continually finishes early. Nonetheless, in the short run we believe:

Proposition 6:   *An internal attribution to effort or ability for a positive goal discrepancy leads to early task completion.*

On the other hand, if the PGD is attributed to a task simply being easier than one estimated (perhaps because the person assigning the task and its deadline goal believed it was more difficult than it was), there is no positive feeling generated by finishing early. In those situations, the person will probably just slow down and finish on time.

Proposition 7:   *An internal attribution to misjudgments about task difficulty leads to slowing down and completion of the task on time.*

The person who is assigned the deadline may also make external attributions for a PGD. For example, a task may be removed from one's work agenda or one may find that a portion of some task is already completed. We suspect that the person will again slow down and stretch the available time to meet the deadline under these conditions. This will happen, of course, only when all other things are held constant (e.g., one's progress, pull from another task).

Proposition 8:   *An external attribution for a positive goal discrepancy leads to a slow down and completion of the task on time.*

Spacing While Striving to Reach Deadlines

Spacing is defined as the allocation of resources across tasks. Initially, we expect more time to be allocated to tasks with clear, proximal deadlines. In addition, increases or decreases in resources allocated to a particular task as a result of pacing may produce reallocations across tasks. However, the most extreme form of spacing is called switching. We stop working on task A and start working on task B. Thus, switching involves a marked qualitative shift in attention and a shift of resources. The most prominent and well accepted psychological mechanism for why people change from one activity to another utilizes an expected value formulation (Atkinson & Birch, 1970). When a set of behaviors (designed to meet a particular deadline) is thought to result in more positive outcomes than what one is currently doing, a shift in the dominant tendency occurs (Atkinson & Birch, 1970; Nuttin, 1985). Naylor, Pritchard, and Ilgen (1980) use a similar, utility-type analysis, which "is intended to explain choices among acts as competing options for resource allocation" (Naylor & Ilgen, 1984, p. 110). More recent work in action theory (Heckheusen, 1991; Sorrentino, 1996) control theory (Kernan & Lord, 1990), and goal setting (Philips et al., 1996) suggests a similar mechanism.

Two obvious situations that will cause a switch in focus are when a task is completed or when one gives up on meeting a particular deadline (Ford, 1992). In addition, a positive goal discrepancy can lead to a task switch when we see we are way ahead of schedule on a task. The process that seems to drive these latter switches is an inaccurate or imprecise initial allocation to a particular task or set of tasks, which results in a goal discrepancy. We have already discussed how attributions influence these changes in behavior.

There are other reasons for switching that use an expected value-type analysis. For example, a task may become boring or fatiguing (Cantor & Blanton, 1996; Locke & Kristof, 1996; Sorrentino, 1996). Because boredom and fatigue are negative experiences, another task may become more positively valent, leading to switching (Kernan & Lord, 1990). Interruptions or disruptive events like phone calls or visitors may also cause reassessments of the expected value for reaching a goal (Kanfer & Kanfer, 1991).

Proposition 9:   *Attributions to lack of effort due to fatigue or boredom will result in switching behavior.*

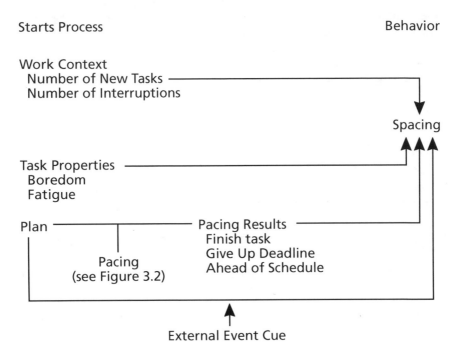

Figure 3.3.   Spacing: Changes in allocation.

One last reason for switching has its roots in a very different psychological mechanism than expected value. A number of authors have recently suggested that task shifts can be built into a daily plan. Bargh and Gollwitzer (1994), for example, describe implementation intentions, which are plans that build in an environmental trigger for a shift in behavior. "Action initiation becomes swift, efficient and does not require conscious thought" (Gollwitzer, 1999, p. 495). In many cases, the task to be started is difficult or negatively valent and thus, the "event" (e.g., the 10:30 AM break, lunch) helps the person get started (Diefendorff & Lord, 2000; Gollwitzer & Schaal, 1998) on a task they might overlook or neglect. Figure 3.3 shows the spacing relationships

CONCLUSION

Our prototypical secretary has multiple tasks and deadlines. In the first stage of our theory, the tasks are judged in terms of their importance, and the deadlines provide goal specificity and goal difficulty information. This information is combined with felt urgency and responsibility to provide an overall commitment to each deadline. A plan is formed that suggests how much effort (usually time) will be allocated to each task (pacing) and across tasks (spacing). Figure 3.1 depicts these relationships.

The second stage of our theory describes how the person strives to meet the deadlines. As work commences, various internal and external events prompt the individual to check their progress; how well their behavior matches the plan. These assessments are indicators of whether the person is on track or is ahead or behind schedule. If on track, the person may hurry up as they approach the goal. If ahead of schedule, they may slow down. When they are behind, they make judgments about whether to hurry up or abandon the deadline. These latter judgments are moderated by task importance, judgments of self-efficacy, causal attributions, and emotional reactions to the NGD. Figure 3.2 shows these relationships.

As work progresses, the individual may shift some resources from one activity to another or actually switch tasks. Some of these switches are caused by the pacing activity. Switches can occur if one is ahead of schedule, finished with a task (meets the deadline), or gives up on a task deadline. Work context variables are also related to switching. New tasks and interruptions should increase switching. The task, if boring or tiring, may produce switches as well. Figure 3.3 shows these relationships.

The contributions of our theory to the research on attributions merit mention. First, attributions are a key process involved in the action cycle. They are an ongoing activity and appear to be crucial to how one actually performs their work assignments. Second, the process is complex. In many cases, the same objective situations (e.g., a specific negative goal discrepancy) will call forth diametrically opposite responses (trying harder vs. giving up) depending on the content of the attribution. Third, the attributions are entwined with other cognitive and emotional reactions. Thus, understanding how attributions influence cognitive variables like self-efficacy and emotional variables like guilt or anger is crucial to enhanced predictions of work behaviors. Finally, we believe that understanding this dynamic and complex process will eventually help us predict (and understand) the slippage between initial goals and plans, and subsequent task performance.

Before closing, however, we want to discuss two areas of attributional research that have not been included in our analysis but certainly could. First, there is currently a lot of research on attributional style (Russell, 1982; Seligman & Schulman, 1986; Thomson & Martinko, Chapter 9, this volume). This work suggests that some people have consistent tendencies over time and tasks to make certain types of attributions. This style could influence how people react to goal discrepancies and their subsequent behavior.

The second area of interest that we did not address has to do with the dynamic nature of the work cycle. Over time, if one continually faces a positive or negative goal discrepancy for a particular type of task, changes are likely to occur in the action cycle. They may set higher or lower goals, adjust their plans (time allocated), increase or decrease their efficacy judgments, and change their attributions. Internal causes may be seen as external, controllable causes as uncontrollable, and so on. While this dynamic process is simply too complex to map at the moment, we hope this chapter provides an initial foundation for such thinking and research.

In conclusion, we have presented a theory that focuses on how the individual adjusts their behavior while working toward deadline goals. We believe a better understanding of attributions and subsequent behaviors like pacing and spacing under conditions of assigned deadlines can be insightful. By analyzing these issues, we can hopefully prescribe ways that individuals can develop effective plans and strategies and work more efficiently as they strive to reach their goals. Also, we can examine the effect of externally and organizationally controlled factors such as the amount of new work and interruptions. By combining better individual strategies and work contexts, we may be able to contribute to the effectiveness of the substantial number of people who work with multiple assigned deadline goals.

REFERENCES

Ajzen, I., & Fishbein, M. (1980). *Understanding attitudes and predicting social behavior.* Englewood Cliffs, NJ: Prentice Hall.

Atkinson, J. W., & Birch, D. (1970). *The dynamics of action.* New York: Wiley.

Austin, J. T., & Bobko, P. (1985). Goal-setting theory: Unexplored areas and future research needs. *Journal of Occupational Psychology, 58,* 289–308.

Austin, J. T., & Vancouver, J. B. (1996). Goal constructs in psychology: Structure, process and content. *Psychological Bulletin, 120,* 338–375.

Bandura, A. (1997). *Self-efficacy: The exercise of control.* New York: Freeman.

Bandura, A., & Cervone, D. (1986). Differential engagement of self-reactive influences in cognitive motivation. *Organizational Behavior and Human Decision Processes, 38,* 92–113.

Bargh, J. A., & Chartrand, J. L. (1999). The unbearable automaticity of being. *American Psychologist, 54,* 462–479.

Bargh, J. A., & Gollwitzer, P. M. (1994). Environmental control of goal-directed action: Automatic and strategic contingencies between situations and behavior. In W. D. Spaulding (Ed.), *Nebraska Symposium on Motivation, Vol. 41: Integrative views of motivation cognition and emotion* (pp. 71–124). Lincoln: University of Nebraska Press.

Blount, S., & Janicek, G.A. (2001). When plans change: Examining how people evaluate timing changes in work organizations. *Academy of Management Review, 26,* 566–585.

Buehler, R., Griffin, D., & Ross, M. (1994). Exploring the planning fallacy: Why people underestimate their task completion time. *Journal of Personality and Social Psychology, 67*, 366–381.

Campion, M. A., & Lord, R. G. (1982). A control systems conceptualization of the goal-setting and changing process. *Organizational Behavior and Human Performance, 30*, 265–287.

Cantor, N., & Blanton, H. (1996). Effortful pursuits of personal goals in daily life. In P. M. Gollwitzer & J. A. Bargh (Eds.), *The psychology of action: Linking cognition and motivation to behavior* (pp. 338–359). New York: Guilford Press.

Carver, C. S., & Scheier, M. F. (1982). Control theory: A useful conceptual framework for personality—social, clinical, and health psychology. *Psychological Bulletin, 92*, 111–135.

Covey, S. R. (1989). *Habits of highly effective people.* New York: Fireside.

Cropanzano, R., James, K., & Citera, M. (1993). A goal hierarchy model of personality, motivation and leadership. *Research in Organizational Behavior, 15*, 267–322.

Cummings, L. L., & Anton, R. J. (1990). The logical and appreciative dimensions of accountability. In S. Sivasta, D. Cooperrider, & Associates (Eds.), *Appreciative management and leadership* (pp. 257–286). San Francisco: Jossey-Bass.

Diefendorff, J. M., & Lord, R. G. (2000, April). *The volitional effects of planning on performance and goal commitment.* Paper presented at the annual meeting of the Society for Industrial and Organizational Psychology, New Orleans, LA.

Donovan J. J., & Swander, C. J. (2001, April). *The impact of self-efficacy, goal commitment, and conscientiousness on goal revision.* Paper presented at the annual meeting of the Society for Industrial and Organizational Psychology, San Diego, CA.

Donovan, J. J., & Williams, K. J. (2003). Missing the mark: Effects of time and causal attributions on goal reversion in response to goal-performance discrepancies. *Journal of Applied Psychology, 88*, 379–390.

Earley, P. C., Shalley, C. E., & Northcraft, G. B. (1992). I think I can, I think I can...Processing time and strategy effects of goal acceptance/rejection decisions. *Organizational Behavior and Human Decision Processes, 53*, 1–13.

Earley, P. C., Wojnaroski, P., & Prest, W. (1987). Task planning and energy expended: Exploration of how goals influence performance. *Journal of Applied Psychology, 77*, 107–114.

Emmons, R. A., & Diener, E. (1886). A goal-effect analysis of everyday situation choices. *Journal of Research in Personality, 20*, 309–326.

Erez, M. (1977). Feedback: A necessary condition for the goal setting–performance relationship. *Journal of Applied Psychology, 62*, 624–627.

Ford, M. E. (1992). *Motivating humans: Goals, emotions and personal agency beliefs.* Newbury Park, CA: Sage.

Frese, M., & Zapf, D. (1994). Action as the core of work psychology; A German approach. In H. C. Triandir, M. D. Dunnette & L. M. Hough (Eds.), *Handbook of industrial and organizational psychology* (2nd ed., Vol. 4, pp. 271–340). Palo Alto, CA: Consulting Psychologists Press.

Frink, D. D., & Klimoski, R. J. (1998). Toward a theory of accountability in organizations and human resources management. *Research on Personnel and Human Resources Management, 16*, 1–51.

Gollwitzer, P. M. (1996). The volitional benefits of planning. In P. M. Gollwitzer & J. A. Bargh (Eds.), *The psychology of action: Linking cognition and motivation to behavior* (pp. 287–312). New York: Guilford Press.

Gollwitzer, P. M. (1999). Implementation intention: Strong effects of simple plans. *American Psychologist, 54,* 493–503.

Gollwitzer, P. M., & Bargh. J. A. (1996). *The psychology of action: Linking cognition and motivation to behavior.* New York: Guilford Press.

Gollwitzer, P. M., & Schaal, B. (1998). Metacognition in action. The importance of implementation intentions. *Personality and Social Psychology Review, 2,* 124–136.

Heckhausen, H. (1991). *Motivation and action.* Berlin: Springer.

Heckhausen, H., & Kuhl, J. (1985). From wishes to action: The dead ends and short cuts on the long way to action. In M. Frese & J. Sabini (Eds.), *Goal directed behavior: The concept of action in psychology* (pp. 134–160). Hillsdale, NJ: Erlbaum.

Heider, F. (1958). *The psychology of interpersonal relations.* New York: Wiley.

Hollenbeck, J. R., & Klein, H. J. (1987). Goal commitment and the goal-setting process: Problems, prospects and proposals for future research. *Journal of Applied Psychology, 72,* 212–220.

Hollenbeck, J. R., & Williams, C. R. (1987). Goal importance, self focus, and the goal setting process. *Journal of Applied Psychology, 72,* 204–211.

Kanfer, R. (1996). Self regulatory and other non-ability determinants of skill acquisition. In P. M. Gollwitzer & J. A. Bargh (Eds.), *The psychology of action: Linking cognition and motivation to behavior* (pp. 404–423). New York: Guilford Press.

Kanfer, R., & Ackerman, P.L. (1989). Motivation and cognitive abilities: An integrative/aptitude-treatment interaction approach to skill acquisition. *Journal of Applied Psychology, 74,* 657–690.

Kanfer, R., & Heggestad, E. D. (1997). Motivational traits and skills: a person-centered approach to work motivation. *Research in Organizational Behavior, 19,* 1–56.

Kanfer, R., & Kanfer F. H. (1991). Goals and self-regulation: Applications of theory to work settings. *Advances in Motivation and Achievement, 7,* 287–326.

Karniol, R., & Ross, M. (1996). The motivational impact of temporal focus: Thinking about the future and the past. *Annual Review of Psychology, 47,* 593–620.

Kelley, H. H. (1967). Attribution in social psychology. *Nebraska Symposium on Motivation, 15,* 192–238.

Kernan, M. C., & Lord, R. G. (1990). Effects of valence, expectancies and goal-performance discrepancies in single and multiple goal environments. *Journal of Applied Psychology, 75,* 194–203.

Klein, H. (1989). An integrated control theory model of work motivation. *Academy of Management Review, 14,* 150–172.

Klein, H. J., & Dineen, B. R. (2002, August). *Predicting changes in goals from goal-performance discrepancies: What's the difference.* Paper presented at the annual meeting of the Academy of Management, Denver, CO.

Klinger, E. (1996). Emotional influences on cognitive processing, with implications for theories of both. In P. M. Gollwitzer & J. A. Bargh (Eds.), *The psychology of action: Linking cognition and motivation to behavior* (pp. 338–359). New York: Guilford Press.

Kluger, A. N., & DeNisi, A. (1996). The effects of feedback interventions of performance: A historical review, a meta-analysis, and a preliminary feedback intervention theory. *Psychological Bulletin, 119*, 259–284.

Kruglanski, A. W. (1996). Goals as knowledge. In P. M. Gollwitzer & J. A. Bargh (Eds.), *The psychology of action: Linking cognition and motivation to behavior* (pp. 599–618). New York: Guilford Press.

Kuhl, J. (1984). Volitional aspects of achievement motivation and learned helplessness: Towards a comprehensive theory of action control. In B. A. Maher (Ed.), *Progress in experimental personality research* (Vol. 13, pp. 99–171). New York: Academic Press.

Kuhl, J. (1996). Who controls whom when "I control myself." *Psychological Inquiry, 7*, 61–68.

Latham, G. P., Erez, M., & Locke, E. A. (1988). Resolving scientific disputes by the joint design of crucial experiments by the antagonists. *Journal of Applied Psychology, 73*, 753–772.

Latham, G. P., & Locke, E. A. (1991). Self-regulation through goal setting. *Organizational Behavior and Human Decision Processes, 50*, 212–247.

Latham, G. P., & Saari, L. M. (1982). The importance of union acceptance for productivity improvement through goal setting. *Personnel Psychology, 35*, 781–787.

Lawrence, T. B., Winn, M. I., & Jennings, P. D. (2001). The temporal dynamics of institutionalization. *Academy of Management Review, 26*, 621–644.

Lim, S. G.-S., & Murnighan, J. K. (1994). Phases, deadlines and the bargaining process. *Organizational Behavior and Human Decision Processes, 58*, 153–171.

Locke, E. A. (1994). The emperor is naked. *Applied Psychology: An International Review, 43*, 367–372.

Locke, E. A., & Kristof, A. L. (1996). Motivational choices in the goal achievement process. In P. M. Gollwitzer & J. A. Bargh (Eds.), *The psychology of action: Linking cognition and motivation to behavior* (pp. 365–384). New York: Guilford Press.

Locke, E. A., & Latham, G. P. (1990). *A theory of goal setting and task performance.* Englewood Cliffs, NJ: Prentice Hall.

Lord, R. G., & Hanges, P. J. (1987). A control system model of organizational motivation: Theoretical development and applied implications. *Behavioral Science, 32*, 161–178.

Lord, R. G., & Levy, P. E. (1994). Moving from cognition to action: A control theory perspective. *Applied Psychology: An International Review, 43*, 335–367.

Markman, A. B., & Brendl, C. M. (2000). The influence of goals on value and choice. *The Psychology of Learning and Motivation, 39*, 97–128.

Martin, L. L., Tesser, A., & McIntosh, W. D. (1993). Wanting but not having: The effects of unattained goals on thoughts and feelings. In D. M. Wegner & J. W. Pennebaker (Eds.), *Handbook of mental control* (pp. 552–572). Englewood Cliffs, NJ: Prentice Hall.

McGrath, J. E., & Rotchford, N. W. (1983). Time and behavior in organizations. *Research in Organizational Behavior, 5*, 57–102.

Miller, G. A., Galanter, E., & Pribraum, K. H. (1960). *Plans and the structure of behavior.* New York: Holt, Rinehart & Winston.

Naylor, J. C., & Ilgen, D. R. (1984). Goal setting: A theoretical analysis of a motivational technology. *Research in Organizational Behavior, 6*, 95–140.

Naylor, J. C., Pritchard, R. D., & Ilgen, D. R. (1980). *A theory of behavior in organizations.* New York: Academic Press.

Nuttin, J. (1985). *Future time perspective and motivation.* Hillsdale, NJ: Erlbaum.

Powers, W. T. (1973). *Behavior: The control of perception.* Chicago: Aldine.

Russell, D. (1982). The causal dimensions scale: A measure of how individuals perceive causes. *Journal of Personality and Social Psychology, 42,* 1137–1145.

Seligman, M. E. P. (1986). Explanatory style as a predictor of productivity and quitting among life insurance agents. *Journal of Personality and Social Psychology, 50,* 832–838.

Seligman, M. E. P. (1993). *Helplessness: On depression, development and death.* San Francisco: Freeman.

Seligman, M. E., & Schulman, P. (1986). Explanatory style as a predictor of productivity and quitting among life insurance sales agents. *Journal of Personality and Social Psychology, 50*(4), 832–838.

Smith, P. C., & Lem, G. (1955). Positive aspects of motivation in repetitive work: Effects of lot size upon spacing of voluntary work stoppages. *Journal of Applied Psychology, 39,* 330–333.

Sorrentino, R. M. (1996). The role of conscious thought in a theory of motivation and cognition. In P. M. Gollwitzer & J. A. Bargh (Eds.), *The psychology of action: Linking cognition and motivation to behavior* (pp. 619–644). New York: Guilford Press.

Staw, B. M. (1976). Knee-deep in the big muddy: A study of escalating commitment to a chosen course of action. *Organizational Behavior and Human Performance, 16,* 27–44.

Thomas, K. M., & Mathiew, J. E. (1994). Role of causal attributions in dynamic self-regulation and goal processes. *Journal of Applied Psychology, 79,* 812–818.

Tubbs, M. E., & Ekeberg, S. E. (1991). The role of intentions in work motivation: Implications for goal setting theory and research. *Academy of Management Review, 16,* 180–199.

Waller, M. J., Conte, J. M., Gibson, C. B., & Carpenter, M. A. (2001). The effect of individual perceptions of deadlines on team performance. *Academy of Management Review, 26,* 586–600.

Weiner, B. (1977). Attribution and affect: Comment on Sohn's Critique. *Journal of Educational Psychology, 69,* 506–511.

Weiner, B. (1985). An attributional theory of achievement motivation and emotion. *Psychological Review, 92,* 548–573.

Weiner, B. (1990). Attribution in personality psychology. In L. A. Perrin (Ed.), *Handbook of personality: Theory and research* (pp. 465–485). New York: Guilford Press.

Weiner, B., & Graham, S. (1999). Attribution in personality psychology. In L. A. Perrin & O. P. John (Eds.), *Handbook of personality: Theory and research* (2nd ed., pp. 605–628). New York: Guilford Press.

Williams, K. J., Donovan, J. J., & Dodge, T. L. (2000). Self-regulation of performance: Goal establishment and goal revision processes in athletes. *Human Performance, 13,* 159–180.

CHAPTER 4

POSITIVE AND NEGATIVE AFFECT AND EXPLANATORY STYLE AS PREDICTORS OF WORK ATTITUDES[1]

William L. Gardner
University of Nebraska–Lincoln

Elizabeth J. Rozell
Southwest Missouri State University

Fred O. Walumbwa
University of Nebraska–Lincoln

ABSTRACT

Using a sample of 205 manufacturing employees, the interrelationships of dispositional affectivity, explanatory style, and work attitudes were explored. Structural equation modeling (SEM) revealed that positive affect (PA) was positively related to more optimistic explanatory styles, job satisfaction, and organizational commitment, while negative affect (NA) was negatively related to employee optimism and job satisfaction. Contrary to expectations, optimism was negatively related to job satisfaction and organizational commitment. A positive relationship between job satisfaction and affective com-

Attribution Theory in the Organizational Sciences, pages 49–81
Copyright © 2004 by Information Age Publishing
All rights of reproduction in any form reserved.

mitment was also identified, and these variables were negatively related to turnover intentions. Finally, dispositional affectivity and explanatory style were indirectly related to work attitudes through mediating variables.

Are persons who tend to be happy also optimistic? How about those who are predisposed to experience negative emotions—do they tend to be pessimistic? And how do such positive and negative patterns of affect and thought relate to work attitudes? The study of work attitudes such as job satisfaction (Herzberg, 1966; Judge, Bono, Thoresen, & Patton, 2003; Locke, 1976; Roethlisberger & Dickson, 1939; Smith, Kendall, & Hulin, 1969), organizational commitment (Cohen, 1993; Mathieu & Zajac, 1990; Mowday, Porter, & Steers, 1982), turnover intentions (Price, 1977; Steers & Mowday, 1981; Tett & Meyer, 1993), and their effects on individual and organizational outcomes, has a long and influential history within the field of organizational behavior (Wren, 1994). Over the past two decades, the focus for much of this research has shifted to the influence that individual difference variables exert on such outcomes (Cropanzano, James, & Konovsky, 1993). However, the interrelationships of people's dispositional tendencies to experience positive affect (PA) and negative affect (NA) (Diener & Emmons, 1985; Watson & Tellegen, 1985), and their tendencies to invoke optimistic versus pessimistic explanations for positive and negative events (Martinko, 2002; Seligman, 1990, 2002), as well as their joint effects on work attitudes, have yet to be explored. This is true despite recent calls from the positive psychology (Cameron, Dutton, & Quinn, 2003; Seligman, 2002; Seligman & Csikszentmihalyi, 2000; Snyder & Lopez, 2002) and the positive organizational behavior (POB; Luthans, 2002a, 2002b) movements to focus greater scholarly attention on positive psychological strengths, such as optimism and subjective well-being (i.e., happiness).

This gap in the literature is surprising for two reasons. First, dispositional affectivity has obvious and important implications for the study of work attitudes, which involve, after all, people's affective reactions to their work (Judge, 1992; Locke, 1976). Importantly, empirical research has confirmed the utility of PA and/or NA in predicting the work outcomes of job satisfaction (Cropanzano et al., 1993; Duffy & Shaw, 1998; Levin & Stokes, 1989; Paradowski, 2001; Pelled & Xin, 1999; Schaubroeck, Judge, & Taylor, 1998; Shaw, 1999; Shaw, Duffy, Abdulla, & Singh, 2000; Shaw, Duffy, Jenkins, & Gupta, 1999; Staw, Bell, & Clausen, 1986; Staw & Ross, 1985; Steel & Rentsch, 1997), organizational commitment (Cropanzano et al., 1993), and turnover intentions (Cropanzano et al., 1993; Duffy & Shaw, 1998; Judge, 1993; Pelled & Xin, 1999; Shaw, 1999; Shaw et al., 2000).

Second, enduring explanatory styles are posited to predispose individuals to make either optimistic or pessimistic attributions for positive and negative events that, in turn, impact their expectations for future outcomes and affective states (Campbell & Martinko, 1998; Gundlach, Douglas, & Martinko, 2003; Martinko, 1995, 2002; Martinko, Gundlach, & Douglas, 2002; Peterson, Maier, & Seligman, 1993; Peterson & Seligman, 1984; Seligman, 1990, 2002)—including their work attitudes. Empirical support for this prediction is provided by Furnham and colleagues (1992), who identified relationships between explanatory style and the attitudinal variables of job satisfaction and intrinsic work motivation.

The purpose of the present study is to take an initial step toward filling this void. We begin by reviewing the pertinent literature for the individual difference variables of dispositional affectivity and explanatory style. Next, we consider the conceptual linkages between these variables, as well as their effects on the work attitudes of job satisfaction, organizational commitment, and turnover intentions. Drawing on this discussion, we use a structural equations methodology (SEM) to test a set of hypotheses regarding the relationships between dispositional affectivity, explanatory style, and certain work attitudes, while comparing alternative conceptual models of these relationships. We examine these models using a sample of 205 employees of two midwestern manufacturing companies. We conclude with a discussion of the results and their implications for management research and practice.

LITERATURE REVIEW

Dispositional Affectivity

Social scientists have long been intrigued by individual differences in people's interpretation of their own emotional experience (Berry & Hansen, 1996). For example, some individuals appear to be predisposed to experiencing higher levels of PA across situations than do others. Such high PA individuals are usually self-described as joyful, exhilarated, excited, and enthusiastic. In contrast, other persons are predisposed to experience more negative emotions than others (Berry & Hansen, 1996; Cropanzano et al., 1993; DePaoli & Sweeny, 2000). High NA individuals often report being afraid, anxious, and angry, and tend to be nervous and tense. As research into dispositional affectivity expanded, the cumulative evidence came to suggest that trait-positive affect (PA) and trait-negative affect (NA) represent two general dimensions of affective responding. Importantly, these dimensions do not appear to represent opposite ends of a continuum; instead, they are independent of one another (Berry & Hansen, 1996; Diener & Emmons, 1985). That is, it is possible for an individual to be high on both, low on both, or high on one but not the other (George,

1992; Watson & Tellegen, 1985). An individual who rates high on both dimensions would be characterized as quite emotional, and would experience fluctuating moods in response to environmental stimuli (Diener & Emmons, 1985). In sharp contrast, an individual who rates low on both would most likely display little affect; that is, the person would be unemotional and unresponsive (Cropanzano et al., 1993).

Interest in dispositional affectivity as a predictor of work attitudes has grown in the wake of empirical evidence that some individuals may be dispositionally predisposed to form positive or negative attitudes about their work. For instance, in a study of 34 monozygotic twin pairs who had been raised apart, Arvey, Bouchard, Segal, and Abraham (1989) demonstrated that approximately 30% of the observed variance in general job satisfaction was attributable to genetic factors. Furthermore, longitudinal studies indicate that scores on job satisfaction measures remain correlated over time, and that this relationship holds even when individuals change employers or occupations (Staw et al., 1986; Staw & Ross, 1985). As Newton and Keenan (1991) and Cropanzano and colleagues (1993) point out, these findings do not mean that work attitudes are entirely stable, or that the job context is unimportant; in actuality, work attitudes do indeed vary over time. Instead, these longitudinal studies are consistent with the view that while work attitudes vary as a function of changes in the work setting, the rank ordering of individuals' attitudes remain relatively stable, and that such stability can be attributed to certain underlying personality dispositions (Cropanzano et al., 1993; George, 1992).

Initially the focus of this stream of research was limited to demonstrating the existence of work attitude predispositions; however, it quickly shifted to the specific personality traits that underlie such predispositions (Cropanzano et al., 1993). Toward this end, trait PA and NA immediately emerged as obvious traits to consider since work attitudes reflect affective reactions to one's work (Cropanzano et al., 1993; Locke, 1976). Several researchers have documented the relevance of dispositional affectivity to work attitudes. For example, an inverse relationship has been found to exist between NA and job satisfaction (Cropanzano et al., 1993; Levin & Stokes, 1989; Paradowski, 2001; Schaubroeck et al., 1998; Shaw et al., 1999; Staw et al., 1986). Other researchers have shown that NA may be negatively correlated with not only job satisfaction, but also organizational commitment, and positively correlated with turnover intentions; the opposite pattern of correlations has been obtained for PA (Cropanzano et al., 1993). One explanation for these relationships is that work attitudes are primarily a function of how one affectively responds to one's work environment, and are therefore influenced by one's underlying affective disposition. Consequently, high PA individuals are likely to exhibit extremely positive responses to their work environment, which are reflected in their work attitudes, while extremely negative responses are usually seen in high NA persons (George, 1992).

Explanatory Style

Paralleling this research has been the development of Martin E. P. Seligman's (1990, 2002) theory of learned optimism. This perspective is rooted in the attributional theory of learned helplessness originally posited by Seligman and his colleagues (Abramson, Seligman, & Teasdale, 1978; Peterson et al., 1993). Their central thesis is that individuals may "learn to be helpless" as a result of repeated exposures to uncontrollable events that produce motivational (passive behavior), cognitive (impaired judgment), and emotional deficits (e.g., depression, shame, anxiety, and hostility). Such deficits are most likely when the uncontrollable events are adverse and attributed by the actor to internal, stable, and global causes (e.g., lack of ability).

In their later works, Seligman and his associates (Buchanan & Seligman, 1995; Seligman, 1990, 2002) posit that individuals possess enduring explanatory styles that predispose them to make either optimistic or pessimistic attributions for positive and negative events that, in turn, determine their expectations for future outcomes and affective state. Persons possessing optimistic explanatory styles are predisposed to attribute positive events to internal, stable, and global causes, while attributing negative events to external, unstable, and specific causes. Persons with pessimistic explanatory styles display the exact opposite pattern of attributions. As a result, they are easily discouraged when confronted with performance obstacles, and are more likely to experience motivational deficits, negative affect, and depression in particular. In contrast, those with optimistic styles are more persistent and resilient, and less likely to experience negative affect.

To operationalize the explanatory style construct, Seligman and his associates developed the Attributional Style Questionnaire (ASQ; Buchanan & Seligman, 1995; Peterson et al., 1982; Seligman, Abramson, Semmel, & von Baeyer, 1979). The 48 items of the ASQ tap various dimensions of the respondents' attributions using six hypothetical positive (success) and six negative (failure) events. ASQ subscales include internality, stability, and globality, CoPos (the composite score for positive events), CoNeg (the composite score for negative events), and CPCN (a total score, the difference between the CoPos and CoNeg scores). The latter three subscales each measure the degree to which a person possesses an optimistic or pessimistic explanatory style. Despite its widespread usage, the reliabilities of the dimensional subscales of the ASQ tend to be low to moderate, with alpha coefficients ranging from .40 to .70. In contrast, the alpha coefficients for the composite scales of CoPos, CoNeg, and CPCN typically exceed .70, and are thus much more satisfactory (Buchanan & Seligman, 1995; Peterson et al., 1993).

To address the psychometric limitations of the ASQ, a variety of alternative instruments have been developed, including the Expanded Attributional Style Questionnaire (EASQ; Peterson & Villanova, 1988), a short form of the EASQ (EASQ-S; Whitley, 1991), and several domain- or respondent-targeted instruments, such as the Academic ASQ (Peterson & Barrett, 1987), the Children's Attributional Style Questionnaire (CASQ; Kaslow, Rehm, Pollack, & Siegel, 1988), the Occupational Attributional Style Questionnaire (OccASQ; Furnham, Sadka, & Brewin, 1992), the Organizational Attributional Style Questionnaire (OrgASQ; Kent & Martinko, 1995), and a revised version of the Organizational Attributional Style Questionnaire (OrgASQ-2; Martinko, 1998). Unfortunately, the reliabilites of the dimensional subscales for many of these alternative instruments are also suspect (Buchanan & Seligman, 1995; Martinko, 1995; Peterson et al., 1993; Peterson & Villanova, 1988). As such, it is common convention to exclusively use the composite scores as indicators of explanatory style (Buchanan & Seligman, 1995).

Research using the ASQ and the alternative measures confirm that people with pessimistic as opposed to optimistic explanatory styles are far more susceptible to cognitive, motivational, and affective deficits. For example, Seligman and Schulman (1986) demonstrated that insurance agents with optimistic as opposed to pessimistic styles performed at higher levels in terms of insurance sales, and exhibited lower levels of turnover. Seligman has also demonstrated that explanatory style is a strong predictor of an individual's affective state, and depression in particular, as predicted (Peterson et al., 1993; Peterson & Seligman, 1984; Seligman, 1990; Seligman et al., 1979). Furthermore, he asserts that it is an important determinant of work attitudes, with optimistic individuals exhibiting considerably more positive attitudes than their pessimistic counterparts (Seligman, 1990, 2002). Support for this argument is provided by Furnham and colleagues (1992), who identified several relationships between explanatory style and the attitudinal variables of job satisfaction and intrinsic work motivation that are consistent with Seligman's theory. Finally, it should be noted that Schulman, Keith, and Seligman (1993) present evidence that optimism and pessimism are heritable by demonstrating that intraclass correlations on the ASQ for monozygotic (e.g., identical) as opposed to dizygotic (fraternal) twin pairs are .48 and 0, respectively. Thus it appears that the origins of explanatory style, like dispositional affectivity, may possess a genetic component. Other factors that are expected to influence the development of one's attributional style include modeling, teachers' differential

feedback, differential exposure to noncontingency, and trust in close relationships (Eisner, 1995). Importantly, both the positive psychology (Seligman, 1990, 2002) and the related POB (Luthans, 2002a, 2002b)
movements posit, and empirical evidence documents (Brockner & Guare,
1983; Fosterling, 1985), that persons who are predisposed to make pessimistic attributions can learn to adopt more optimistic explanations
through attributional training.

RESEARCH MODEL HYPOTHESES

From the preceding discussion of the dispositional affectivity and explanatory style literatures, several areas of conceptual overlap between these and
related constructs are apparent. Specifically, in each perspective, the core
construct is viewed as a relatively stable characteristic of the individual,
which is either composed of, or related to, affect, and which has been
shown to be predictive of certain work attitudes. In light of the apparent
linkages between these constructs, it is somewhat surprising to find that the
relationships between them, as well as their combined effects on work attitudes, have yet to be explored.

   Toward this end, this research explored the interrelationships among
these constructs as reflected in the research model depicted in Figure 4.1.
As the model indicates, dispositional affectivity and explanatory style are
posited to be interrelated antecedents of work attitudes. That is, these

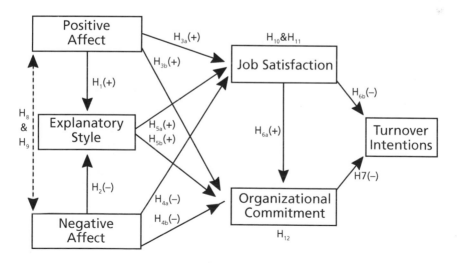

Figure 4.1.   Direct and indirect effects of dispositional affectivity on work-related
attitudes and behaviors.

constructs are considered to be relatively permanent attributes of the individual that account for variance in emergent job attitudes.

While Seligman (1990, 2002) asserts that the manner in which one habitually explains positive and negative events accounts for one's affective tendencies, an equally compelling argument can be made for an opposite causal effect, whereby affective disposition shapes one's explanations for life events (Carver & Scheier, 1990). Here, it is worth noting that in interpreting the results of their twins study, Schulman and colleagues (1993) speculate that explanatory style may be only indirectly heritable, and merely correlated with another trait that is directly heritable. Specifically, they posit that such heritable molar traits as intelligence, temperament, attractiveness, and coordination impact the amount of success or failure a person experiences, which may in turn influence explanatory style. Importantly, the notion of temperament as used by Schulman and colleagues is similar to the construct of dispositional affectivity. Accordingly, we offer a tentative extension of their argument by positing that dispositional affectivity constitutes a heritable and stable trait that impacts the level of success and failure that individuals experience; such life experiences, in turn, shape the development of their explanatory styles.

The antecedents of dispositional affectivity and explanatory style are, in turn, posited to directly influence job satisfaction and organizational commitment, as depicted in Figure 4.1. Given our focus on affect, we limit our attention to the affective (affective attachment to the organization), as opposed to the continuance (the perceived cost of leaving the organization) and normative (perceived obligation to remain with the organization) components of organizational commitment identified by Meyer and Allen (1991). An indirect relationship between dispositional affectivity and these work attitudes, as mediated by explanatory style, is also posited. Job satisfaction and affective commitment are likewise expected to mediate the relationships between these antecedents and turnover intentions; support for this assertion is available from Cropanzano and colleagues (1993) and Robert, Probst, Martocchio, Drasgow, and Lawler (2000). While some controversy exists about the direction of the relationship between job satisfaction and organizational commitment (Martin & Bennett, 1996; Williams & Hazer, 1986), the weight of the conceptual and empirical evidence suggests that job satisfaction contributes to the development of organizational commitment, rather than vice versa (Robert et al., 2000; Williams & Hazer, 1986). Based on this reasoning, we advance the hypotheses depicted in Figure 4.1 and listed below.

Hypothesis 1:   *PA is positively related to the extent to which employees exhibit more optimistic (and less pessimistic) explanatory styles.*

Hypothesis 2:   *NA is negatively related to the extent to which employees exhibit more optimistic (and less pessimistic) explanatory styles.*

Hypothesis 3:   *PA is positively related to the extent to which employees experience higher levels of (a) job satisfaction and (b) affective commitment.*

Hypothesis 4:   *NA is negatively related to the extent to which employees experience higher levels of (a) job satisfaction and (b) affective commitment.*

Hypothesis 5:   *The extent to which employees exhibit more optimistic (and less pessimistic) explanatory styles is positively related to (a) job satisfaction and (b) affective commitment.*

Hypothesis 6:   *Employee job satisfaction is (a) positively related to affective commitment and (b) negatively related to turnover intentions.*

Hypothesis 7:   *Affective commitment is negatively related to employees' turnover intentions.*

Hypothesis 8:   *Explanatory style mediates the relationships between PA and the levels of (a) job satisfaction and (b) affective commitment experienced by employees.*

Hypothesis 9:   *Explanatory style mediates the relationships between NA and the levels of (a) job satisfaction and (b) affective commitment experienced by employees.*

Hypothesis 10:   *Employee job satisfaction mediates the relationships of (a) PA, (b) NA, and (c) explanatory style with affective commitment.*

Hypothesis 11:   *Employee job satisfaction mediates the relationships of (a) PA, (b) NA, and (c) explanatory style with turnover intentions.*

Hypothesis 12:   *Affective commitment mediates the relationships of (a) PA, (b) NA, (c) explanatory style and (d) job satisfaction with turnover intentions.*

METHOD

Participants

The population consisted of 685 salaried and hourly employees of two manufacturing firms located in the midwestern United States. Of the survey packets distributed, 205 were completed and returned for a response rate of 30%. Table 4.1 provides a summary of the demographic attributes of the participants.

Measures

A variety of instruments were used to measure the variables included in the research model. Descriptions of these measures and their psychometric properties are provided below.

Explanatory style. As noted above, a wide variety of explanatory style measures have been developed over the years. In choosing a measure for this study, three key criteria were established. First, it had to include scenarios reflecting both positive and negative events since differential relationships between PA and NA and the CoPos and CoNeg scales may exist that are of interest. Second, evidence of acceptable reliabilities for both the dimensional subscales and the composite scales was required. Third, the instrument had to include procedures and scales that were comparable to the ASQ in order to facilitate the interpretation of the results and comparisons to those of prior studies. Unfortunately, none of the available measures adequately satisfied all of these criteria. For this reason, we selected the original ASQ, since it is the most extensively validated and widely used instrument. In an effort to improve the reliabilities of the dimensional subscales, however, we added two positive and two negative scenarios that were selected from the EASQ and Furnham and colleagues'

Table 4.1. Demographic Attributes of the Subjects

Attribute	Mean	Std. Dev.
Age	38.5	10.6
Company Tenure	8.1	7.3
Job Tenure	5.3	5.7
Years to Retirement	23.4	10.5

Attribute	Frequency	Percentage
Gender		
Male	88	43.6
Female	114	56.4
Education		
Less Than High School	4	2.0
High School Degree	47	22.9
Some College	49	23.9
AA or 4-Year Degree	81	39.5
Masters	13	6.3
Ph.D.	1	.5
Other	10	4.9

Attribute	Frequency	Percentage
Marital Status		
Single	33	16.2
Married	138	67.6
Widowed	5	24.5
Divorced/Separated	28	13.7

Table 4.1.   Demographic Attributes of the Subjects (Cont.)

Job Status		
Senior/Top Manager	8	3.9
Middle Manager	23	11.3
First Level Manager	22	10.8
Professional	52	25.6
Clerical	62	30.5
Production/Maintenance	36	17.7
Compensation		
Hourly	110	54.5
Salary	92	45.5
Attribute	*Frequency*	*Percentage*
Annual Income		
Less than $10,000	6	3.0
$10,000–$20,000	58	28.6
$20,001–$30,000	62	30.5
$30,001–$40,000	43	21.2
$40,001–$50,000	10	4.9
$50,001–$60,000	8	3.9
$60,001–$80,000	8	3.9
More than $80,000	8	3.9

(1992) OccASQ. This approach is in keeping with that taken by Peterson and Villanova (1988), who limited the EASQ to negative scenarios and greatly expanded the number included from 6 to 24 in an effort to improve subscale reliabilities.

The resultant measure includes the 12 original scenarios of the ASQ (six positive and six negative), plus four supplemental scenarios. Sample positive scenarios include "You do a project which is highly praised" and "You are given a special performance award at work." Sample negative scenarios include "You have been looking for a job unsuccessfully for some time" and "You give an important talk in front of a group and the audience reacts negatively." The instructions, procedures, and scales of the ASQ were retained. For each scenario, respondents are instructed to: (1) vividly imagine themselves in the situations described; (2) decide what they feel would be the major cause of the situation if it happened to them; (3) record this cause in the blank space provided; (4) respond to the three questions measuring internality, stability, and globality; and (5) go on to the next question.

In the current study, coefficient alphas of .77, .79, .56, .70, and .72, respectively, were obtained for the CoPos and CoNeg, internality, stability, and globality subscales. Thus, as in prior studies (Peterson et al., 1982,

1993), acceptable levels of reliability were obtained for all of the subscales with the exception of internality. In keeping with convention, the overall composite scale (CPCN) was used to test the proposed research model and hypotheses.

Dispositional affectivity. PA and NA were measured using the Positive and Negative Affect Schedule (PANAS) developed by Watson, Clark, and Tellegen (1988). The PANAS includes a list of 20 mood-relevant adjectives, of which 10 indicate positive (e.g., active, enthusiastic) and 10 indicate negative (e.g., angry, afraid) moods. Although both trait and state instructions are available for the PANAS (Schmukle, Egloff, & Burns, 2002), trait instructions were used in this study. Specifically, respondents were instructed to "indicate to what extent you generally feel this way, that is, how you feel on the average." Extensive validity evidence for the PANAS is available from Watson, Clark, and Tellegen (1988), Watson, Clark, and Carey (1988), Watson (1988a, 1988b), and DePaoli and Sweeny (2000). Alpha coefficients for the PA scale range from .86 to .90, while those of the NA scale range from .84 to .87, and are thus acceptable. Similarly, coefficient alphas of .84 and .85 for the PA and NA scales, respectively, were obtained in the current study.

Job satisfaction. Overall job satisfaction was measured using the 18-item "Job in General" (JIG) scale (Ironson, Smith, Brannick, Gibson, & Paul, 1989) from the revised version of the Job Descriptive Index (JDI) (Smith et al., 1969). The JIG includes 18 descriptive adjectives (e.g., pleasant, bad, waste of time, worthwhile) that respondents use to rate their job. Specifically, respondents record a Y for "yes" or an N for "no" to indicate if the statement does or does not describe their job, respectively, and a "?" if they cannot decide. Validation evidence for the JIG is available from Ironson and colleagues (1989); coefficient alphas for the JIG scale range from .91 to .95. In the present study, an alpha coefficient of .89 was obtained.

Organizational commitment. In a review of the organizational commitment literature, Meyer and Allen (1991) identified *affective, continuance,* and *normative* commitment as three distinct components of commitment. Because the standard Organizational Commitment Questionnaire (Mowday et al., 1982) fails to distinguish between these components, Allen and Meyer (1990) developed and validated separate measures for each component. Given our focus on affective commitment, we included their eight-item Affective Commitment Scale (ACS) as our measure of organizational commitment. Sample items include "I would be happy to spend the rest of my career with this company" and "I do not feel like 'part of the family' at my company" (reverse scored). This measure uses a 7-point response scale with the following anchors: 1 = strongly disagree; 2 = moderately disagree; 3 = slightly disagree; 4 = neither agree nor disagree; 5 = slightly agree; 6 = moderately agree; and 7 = strongly agree. Coefficient alphas for the ACS of

.87 and .89 were obtained by Allen and Meyer, and in the present study, respectively.

Turnover intentions. A measure of intent to leave developed by O'Reilly, Chatman, and Caldwell (1991) was employed in this study. This scale is composed of four 7-point Likert-type questions: (1) "To what extent would you prefer another more ideal job than the one you now work in?"; (2) "To what extent have you thought seriously about changing organizations since beginning to work here?"; (3) "How long do you intend to remain with this organization?"; and (4) "If you have your own way, will you be working for this organization three years from now?" We obtained a coefficient alpha of .85 for this scale.

Procedure

The instrument packets were administered in cooperation with contact members of the targeted organizations. Specifically, convenience sampling was employed whereby the contact persons distributed the instrument packets to all hourly and salaried employees in their work units. Respondents completed the instruments during normal work hours, and returned them directly to the researchers using a preaddressed and prepaid postage packet.

Tests of Alternative Models

Because SEM only provides information on the goodness of fit for a proposed model, and cannot be used to determine if such a model is the "correct" one, it is important to examine other theoretically plausible models. Toward this end, we compared the theoretical model depicted in Figure 4.1 with several alternative models. The distinctions between these models stem from: (a) the measures used to operationalize explanatory style (CoNeg, CoPos, and CPCN) and (b) the causal direction posited between the dispositional affectivity and explanatory style variables.

As indicated during our discussion of the ASQ, the CoPos, CoNeg, and CPCN scales each serve as measures of explanatory style. An argument can be made that PA and NA will relate separately and directly to the CoPos and CoNeg measures, respectively, rather than to the composite CPCN measure. That is, PA appears to be conceptually linked to CoPos, which measures one's cognitive explanations for positive events, while NA is more relevant to the CoNeg scale, which assesses one's explanations for negative events. Accordingly, we tested a model in which PA and NA are posited to exert direct effects on job satisfaction and affective commitment, as well as mediated effects through CoPos and CoNeg, respectively.

For both the separate and the combined measures of explanatory style, we also tested alternative models in which the predicted relationships with dispositional affectivity are reversed; that is, explanatory style is posited to be an antecedent for affectivity, rather than vice versa. Such a reversal is conceptually viable and consistent with Seligman's (1990, 2002) assertion that explanatory style is a determinant of one's attributions for success and failure, and subsequent affective reactions. Support for such a model would suggest that, in this study, the PANAS served as a measure of state, as opposed to trait, affectivity (Schmukle et al., 2002).

The results of these analyses revealed that, when compared to the alternative models, the proposed model that included the composite CPCN scale yielded the best fit and predictive estimates. These findings are consistent with prior studies (e.g., Seligman & Schulman, 1986), which found the CPCN scale to be a stronger predictor of work outcomes than the separate CoPos and CoNeg scales. The alternative model that included CoPos and CoNeg separately did reveal that PA and NA, respectively, were positively and significantly related to these scales, as expected, whereas the relationships of NA with CoPos and PA with CoNeg were insignificant. However, while CoNeg emerged as a significant predictor of work attitudes, CoPos did not. Moreover, fit indices indicated that neither this nor the alternative models in which the causal relationships between dispositional affectivity and explanatory style were reversed fit the data as well as the proposed model. Due to space considerations, the tests of the alternative models are not reported here. Instead, we provide a detailed discussion of the results obtained for the proposed model and research hypotheses below.

RESULTS

Descriptive Statistics

A summary of the means, standard deviations, and intercorrelations for the focal variables is provided in Table 4.2. These correlations are consistent with the underlying theory and prior research, providing evidence of convergent and discriminant validity. For example, PA and NA are negatively, but not strongly, correlated, providing support for the notion that these are related but independent constructs. Similarly, the relationships between job satisfaction, affective commitment, and turnover intentions are in the expected directions and consistent with prior research (Cropanzano et al., 1993; Williams & Hazer, 1986).

Measurement Model

Before testing the various aspects of the proposed theoretical model, we first assessed the suitability of the measurement model using AMOS maximum likelihood estimation procedure and assuming covariance (Arbuckle & Wothke, 1999). A measurement model provides the link between scores on a measuring instrument and the underlying constructs they are designed to measure (Byrne, 2001b). To do this, we formed two to four multi-item indicators for each of the constructs. This procedure was preferred because it minimizes the extent to which the indicators of each construct share variance, in addition to generating more stable parameter estimates (Fitzgerald, Drasgow, Hulin, Gelfand, & Magley, 1997).

Table 4.2.   Variables Means, Standard Deviations, and Intercorrelations [a]

Variable	Mean	Std. Dev.	1	2	3	4	5
1. Positive Affect	3.68	.58					
2. Negative Affect	1.79	.59	−.23**				
3. Explanatory Style (CPCN)	1.15	.83	.17*	−.22**			
4. Job Satisfaction	2.41	.54	.30**	−.42**	−.03		
5. Affective Organizational Commitment	4.01	1.21	.21**	−.16**	−.12	.53**	
6. Turnover Intentions	3.18	1.42	−.19**	.19**	.07	−.59**	−.41**

[a] $n = 205$; * $p < .05$; ** $p < .01$

In assessing the proposed model, we relied on several standard fit indices to examine the overall model fit. These included the (a) change in the chi-square relative to the change in degrees of freedom; (b) ratio of chi-square to degrees of freedom; (c) goodness-of-fit index (GFI); (d) comparative fit index (CFI); and (e) root mean square error of approximation (RMSEA). The chi-square/degrees of freedom ratio was 1.23, the GFI was .93, the CFI was .98, and the RMSEA was .03, indicating a good fit. The chi-square and degrees of freedom for the overall model were 144.32 and 117, respectively. In addition, all the estimated factor loadings were significant and reasonably close to 1.00, suggesting that the indicators measured the latent traits well.

Tests of Hypotheses

Having established the invariant structure of the theoretical model, we performed a series of path analyses using AMOS maximum likelihood estimation procedure (Arbuckle & Wothke, 1999). Path analysis was used because it allows for the examination of the direct or indirect influence that a particular latent variable has on other latent variables within the model (Byrne, 2001a, 2001b). The fit indices for the structural model were: chi-square = 134.10, degrees of freedom = 119, chi-square/degrees of freedom ratio = 1.13, GFI = .93, CFI = .99, and RMSEA = .03, indicating a very good fit. The results of the path analysis are shown in Figure 4.2.

As the figure indicates, the relationships of PA with explanatory style (CPCN) (standardized β = .24, $p < .01$) and job satisfaction (standardized β = .26, $p < .01$) were significant and positive, providing support for H_1 and H_{3a}, respectively. In addition, the relationship of PA with affective commitment (standardized β = .13, $p < .07$) was positive and marginally significant, providing some support for H_{3b}. As posited by H_2, the relationship of NA with CPCN (standardized β = −.27, $p < .01$) was negative and significant. The negative relationship of NA with job satisfaction predicted by H_{4a} was likewise significant (standardized β = −.44, $p < .001$). However, the expected relationship between NA and affective commitment (H_{4b}) was not significant (standardized β = .06, $p > .10$). Similarly, the relationship between CPCN and affective commitment (H_{5b}) was negative and marginally significant (standardized β = −.17, $p < .06$), but the relationship of CPCN with job satisfaction (H_{5a}) failed to achieve significance (standardized β = −.10, $p > .10$). The predicted relationship of job satisfaction with affective commitment was positive and significant (standardized β = .62, $p < .001$), providing support for H_6. Finally, and consistent with prior research, the relationships of job satisfaction (standardized β = −.59, $p < .001$) and affective commitment (standardized β = −.17, $p < .05$) with employees' turnover intentions were both significant and negative (Robert et al., 2000; Walumbwa & Lawler, 2003); thus, H_{6b} and H_7 were supported.

Mediating Tests

Although the SEM results reported above and depicted in Figure 4.2 provide important insights regarding the study's hypotheses, there are two noteworthy limitations of this analysis. First, the mediating effects posited by Hypotheses 7–12 are not assessed. Second, in lieu of such tests, the assessment of the direct relationships predicted by Hypotheses 1–6 is incomplete and potentially misleading. Thus, it was necessary to supplement the preceding analysis with additional tests for mediating effects. We did so by comparing several variants of the proposed model. The model depicted in Figure 4.1 includes: (a) the *direct* and *indirect* effects of PA and NA on both job satisfaction and affective organizational commitment, (b) the *direct* effects of PA and NA on CPCN, (c) the *direct* effects of CPCN on job satisfaction and affective commitment, (d) the *direct* effects of job satisfaction on affective commitment, (e) the *direct* and *indirect* effects of job satisfaction on turnover intentions, and (f) the *direct* effects of affective commitment on turnover intentions. All of these comparisons were performed using AMOS maximum likelihood (Arbuckle & Wothke, 1999). We provide a summary of the

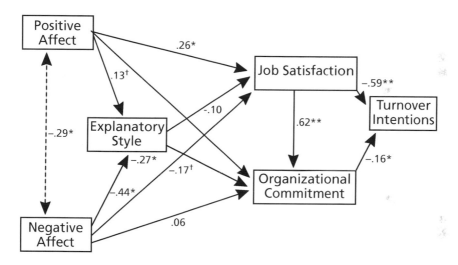

Figure 4.2.   Results of AMOS path analysis (n = 205).
† p < .10; * p < .05; ** p < .01; *** p < .001

different models and the respective path omitted in each model in Table 4.3; note that the top and bottom portions of the model pertain to the effects of PA and NA, respectively.

Table 4.3. Model Comparisons and Results from Path Analyses

	Missing Path	χ^2	df	χ^2/df	GFI	NFI	CFI	Δdf	$\Delta\chi^2$
Positive Affect									
A. Full Model	None	134.10	119	1.13	.93	.93	.99		
B. Full Model without a path from PA to Job Satisfaction	1	145.25	120	1.20	.93	.92	.99	1	11.15***
C. Model B without a path from PA to Affective Commitment	1	136.88	120	1.14	.93	.93	.99	1	8.37***
D. Model B without a path from PA to CPCN	1	142.11	120	1.18	.93	.93	.99	1	3.14†
E. Model B without a path from CPCN to Job Satisfaction	1	135.52	120	1.13	.93	.93	.99	1	9.73**
F. Model B without a path from CPCN to Affective Commitment	1	138.68	120	1.16	.93	.93	.99	1	6.57*
G. Model B without a path from Job Satisfaction to Affective Commitment	1	182.06	120	1.52	.91	.90	.96	1	36.81***
H. Model B without a path from Job Satisfaction to Turnover Intentions	1	174.37	120	1.45	.91	.91	.97	1	29.12***
I. Model B without a path from Affective Commitment to Turnover Intentions	1	137.60	120	1.15	.93	.93	.99	1	7.65**
Negative Affect									
A. Full Model	None	134.10	119	1.13	.93	.93	.99		
J. Full Model without a path from NA to Job Satisfaction	1	163.27	120	1.36	.92	.91	.98	1	29.27***
K. Model J without a path from NA to Affective Commitment	1	134.53	120	1.12	.93	.93	.99	1	28.74***
L. Model J without a path from NA to CPCN	1	144.49	120	1.20	.93	.92	.99	1	18.78***
M. Model J without a path from CPCN to Job Satisfaction	1	135.52	120	1.13	.93	.93	.99	1	27.75***
N. Model J without a path from CPCN to Affective Commitment	1	138.68	120	1.16	.93	.93	.99	1	14.59***
O. Model J without a path from Job Satisfaction to Affective Commitment	1	182.06	120	1.52	.91	.90	.96	1	18.79***
P. Model J without a path from Job Satisfaction to Turnover Intentions	1	174.37	120	1.45	.91	.91	.97	1	11.10***
Q. Model J with a path from Affective Commitment to Turnover Intentions	1	137.60	120	1.15	.93	.93	.99	1	25.67***

† $p < .10$; * $p < .05$; ** $p < .01$; *** $p < .001$

Because Model A shown in Table 4.3 includes all the direct and indirect effects of dispositional affectivity (PA and NA) and explanatory style on work attitudes, it is a saturated model. Model B is similar to Model A, except the direct path from PA to job satisfaction is absent. The chi-square difference between Model A and B is an overall test of the direct effect of PA on job satisfaction (H_{3a}). Model C is similar to Model B, except the direct path from PA to affective commitment is absent. The chi-square difference between Model B and C indicates an overall test for the direct effect of PA on affective commitment (H_{3b}). Model D is similar to Model B, except the path from PA to CPCN is absent. The chi-square difference between Model B and D represents an overall test of the direct effect of PA on CPCN (H_1).

Models E through I examine the direct effects of CPCN and work attitudes, as well as the indirect effects of PA on work attitudes. Models E and F are similar to Model B, except the paths from CPCN to job satisfaction and CPCN to affective commitment are absent, respectively. The chi-square differences between Model B and E and Model B and F, respectively, provide an overall test of the direct effects of CPCN on job satisfaction (H_{5a}) and affective commitment (H_{5b}). Model G is similar to Model B, except the path from job satisfaction to affective commitment is absent. The chi-square difference between Models B and G represents an overall test of the direct effect of job satisfaction on affective commitment (H_{6a}). Finally, Models H and I are similar to Model B, except the paths from job satisfaction and affective commitment to turnover intentions, respectively, are absent. Thus, the chi-square difference between Models B and H and Models B and I, respectively, provide an overall test of the direct effect of job satisfaction (H_{6b}) and affective commitment (H_7) on turnover intentions. In the bottom portion of the table, Models J through L specify the effects of NA that are analogous to those examined by Models B through D for PA; they are used to test H_2, H_{4a} and H_{4b}. In addition, Models M to Q for NA are analogous to Models E to I for PA and provide alternative tests of H_{5a}, H_{5b}, H_{6a}, H_{6b} and H_7.

Support for a direct effect of PA on job satisfaction (H_{3a}) would be observed *if* the chi-square difference between Model A and B is significant. The same would be true for H_{3b} *if* the chi-square difference between Model A and C is significant for affective commitment. On the other hand, it can be argued that PA has an indirect effect mediated through CPCN (H_{8a}) when both the path from PA to CPCN and the path from CPCN to job satisfaction are significant. It also can be argued that PA has an indirect effect on employees' turnover intentions mediated through job satisfaction (H_{9a}) *when* both the path from PA to job satisfaction and a path from job satisfaction to turnover intentions are significant. Similarly, it can be argued that job satisfaction has an indirect effect mediated through affective commitment (H_{11d}) *when* both the path from job satisfaction to affective commitment and the path from affective commitment to turnover intentions are significant. An analogous procedure is used to test the additional mediating effects posited for explanatory style (H_{8b}, H_{9a}, and H_{9b}), job satisfaction (H_{10b}, H_{10c}, H_{11a}, H_{11b}, and H_{11c}), and affective commitment (H_{12a}, H_{12b}, and H_{12c}).

Results of these model comparisons are summarized in Table 4.3. As shown in the table, PA had both direct and indirect effects on job satisfaction and affective commitment, respectively, as well as indirect effects on turnover intentions. The chi-square differences between Models A and B, as well as between Models B and C, were significant, indicating a direct effect of PA on job satisfaction (H_{3a}) and affective commitment (H_{3b}), respectively. Note that the latter relationship did not emerge in the preceding analysis. The chi-square difference between Models B and D was marginally significant, indicating a moderate direct effect of PA on CPCN (H_1). The chi-square difference between Models B and E as well as between Models B and F were statistically significant, indicating an indirect effect of PA on job satisfaction (H_{8a}) and affective commitment (H_{8b}), mediated through CPCN. The chi-square difference between Models B and G was likewise significant, indicating an indirect effect of PA on affective commitment, mediated through job satisfaction (H_{10a}). Similarly, the chi-square difference between Models B and H was significant, indicating an indirect effect of PA on turnover intentions, mediated by job satisfaction (H_{11a}). Finally, the chi-square difference between Models B and I was statistically significant, indicating an indirect effect of PA on turnover intentions, mediated through affective commitment (H_{12a}). Thus, support was obtained for all of the predicted direct and indirect effects of PA.

The bottom half of Table 4.3 summarizes the results of path analyses for NA. As shown in the table, NA also had both direct (H_{4a} and H_{4b}) and indirect effects (H_{9a} and H_{9b}) on job satisfaction and affective commitment, respectively, mediated through CPCN. Note that the direct effects identified did not emerge in the prior analysis, demonstrating the utility of the model comparison procedure. The chi-square differences between Models J and M as well as between Models J and N were statistically significant, indicating an indirect effect of NA on job satisfaction (H_{9a}) and affective commitment (H_{9b}), mediated through CPCN. Similarly, the chi-square difference between Models J and O was significant, suggesting an indirect effect of NA and CPCN on affective commitment, mediated through job satisfaction (H_{10b}). The chi-square difference between Model J and P was likewise statistically significant, suggesting an indirect effect of NA on turnover intentions mediated through job satisfaction (H_{11b}). Finally, the difference between Models J and Q was statistically significant, indicating an indirect effect of NA on turnover intentions, mediated through job satisfaction and affective commitment (H_{12b}). Thus, support was obtained for all of the predicted direct and indirect effects of NA.

Consistent with H_{10c}, our findings suggest that explanatory style is indirectly related to affective commitment through job satisfaction. Similarly, evidence emerged that CPCN exerts indirect effects on turnover intentions through job satisfaction (H_{11c}) and affective commitment (H_{12c}). Support for an indirect relationship of job satisfaction on turnover intentions, mediated by affective commitment (H_{12d}), was likewise obtained. However, contrary to our expectations (H_{5a} and H_{5b}) and the prior analysis, CPCN was inversely related to both job satisfaction and affective commitment. We consider possible explanations for these surprising results below.

DISCUSSION

Recall that the primary objectives of this study were to explore the relationships between dispositional affectivity and explanatory style, as well as their joint effects on the work attitudes of job satisfaction, affective commitment, and turnover intentions. Because of the cross-sectional design, we are unable to identify causal relationships among these variables. Nonetheless, it was possible to examine competing structural models using SEM to determine the model that provided the best fit to the data. Specifically, the proposed theoretical model, in which PA and NA are posited to have direct effects on work attitudes, as well as indirect effects through explanatory style (CPCN), was compared to a plausible alternative model in which the causal relationships of PA and NA with explanatory style were reversed. We also examined competing models that included employees' composite explanations for positive (CoPos) and negative (CoNeg) events as separate constructs that are independently related to PA and NA, respectively. The results of these comparative analyses revealed that the proposed model depicted in Figure 4.1 provided the best fit to the data. Hence, the proposed model was used as the basis for subsequent hypothesis testing. The specific findings obtained from these analyses are discussed below.

Dispositional Affectivity and Explanatory Style

Consistent with H_1 and H_2, PA and NA, respectively, were positively and negatively related to explanatory style (CPCN). These findings suggest that persons who are predisposed to experience PA tend to favor optimistic explanations for the events they encounter. Similarly, trait-NA is associated with a tendency to make pessimistic attributions. Despite conceptual support (e.g., Carver & Scheier, 1990) and suggestive empirical findings (Schulman et al., 1993), these relationships between dispositional affectivity and explanatory style had not been previously established. They imply that dispositional tendencies to experience positive and negative affect influence people's characteristic explanations for positive and negative events, with PA fostering more optimistic explanations, and NA more pessimistic explanations. Of course, the relationship may also work in the opposite causal direction, with explanatory styles impacting the level of PA and NA experienced. Such a relationship would be consistent with Seligman's (1990, 2002) assertion that people who habitually make more pessimistic explanations for negative events will experience heightened levels of NA, whereas those who favor optimistic explanations for positive events will enjoy elevated levels of PA. Note that in this relationship, state-affect, as opposed to trait-affect, is operative, whereas trait instructions for the PANAS were employed in this study. Perhaps this explains why the proposed model provided a better fit to the data. Future research that explores the causal relationships of both trait and state affectivity with explanatory style, as well as work attitudes, is clearly warranted.

Dispositional Affectivity and Work Attitudes

Consistent with H_{3a} and H_{3b}, PA was shown to be significant and positively related to the work attitudes of job satisfaction and affective commitment, respectively. Similarly, the comparative analysis revealed significant and negative relationships of NA with job satisfaction and affective commitment, respectively, providing support for H_{4a} and H_{4b}. These findings imply that employees who are dispositionally inclined to experience PA tend to be more satisfied with their jobs and affectively committed to the organization, while those who are predisposed toward NA have a tendency to be dissatisfied with their jobs and less affectively commited. Importantly, these findings reinforce those of a growing number of studies that demonstrate the utility of PA and/or NA in predicting employee job satisfaction (Cropanzano et al., 1993; Duffy & Shaw, 1998; Levin & Stokes, 1989; Paradowski, 2001; Pelled & Xin, 1999; Schaubroeck et al., 1998; Shaw, 1999; Shaw et al., 1999, 2000; Staw et al., 1986; Steel & Rentsch, 1997), and organizational commitment (Cropanzano et al., 1993). Moreover, the results suggest PA has indirect effects on job satisfaction (H_{9a}) and affective commitment (H_{9b}) through explanatory style, as well as indirect effects on turnover intentions through job satisfaction (H_{11a}) and affective commitment (H_{12a}). Additionally, NA appears to be indirectly related to affective commitment (H_{10b}), as well as turnover intentions (H_{11b}), through job satisfaction. Together, these findings imply that dispositional affectivity has a role to play in explaining why some people are more likely than others to enjoy their jobs, be affectively committed to their organizations, and have intentions to continue their employment.

Explanatory Style and Work Attitudes

Contrary to the predictions of H_{5a} and H_{5b} and our preliminary analysis, our comparative analysis revealed negative relationships of explanatory style with both job satisfaction and affective commitment. In support of Hypotheses 10c and 11c, respectively, evidence that CPCN exerts indirect effects on affective commitment and turnover intentions through job satisfaction was obtained. In addition, an indirect effect of explanatory style on turnover intentions mediated by affective commitment (H_{12c}) was revealed. When considered in conjunction with the mediating effects of CPCN described above, these findings imply that explanatory style plays a central role in determining the levels of job satisfaction and affective commitment that employees experience at work, which, in turn, influence their turnover intentions. Indeed, these results suggest that explanatory style merits increased conceptual and empirical attention as a potential mediator of the relationships between dispositional affectivity and work attitudes.

As for the unexpected negative relationships of explanatory style with job satisfaction and affective commitment, hindsight suggests that they are not completely counterintuitive. Recall that optimistic persons tend to attribute positive events to internal, stable, and global causes such as ability, as opposed to external, unstable, and specific causes (e.g., specific and temporary attributes of the work setting, including help from coworkers or organizational resources). They also tend to attribute negative events to their environment rather than stable and personal attributes. In contrast, pessimists tend to blame themselves for negative events, while attributing positive outcomes to the situation or other persons. Not surprisingly, optimists tend to experience more positive emotions, and fewer negative emotions, than pessimists, as our results confirm. However, because they attribute positive versus negative events to personal rather than situational factors, they may be less likely than pessimists to perceive *their job* as a source of satisfaction; instead, they seem to attribute their satisfaction to personal factors. In contrast, pessimists may view their job as a greater source of satisfaction than they do themselves. The inverse relationship of explanatory style with affective commitment can likewise be explained by the fact that job satisfaction covaries with, and appears to be a causal determinant of, affective commitment. That is, to the extent that optimists are more satisfied with their jobs they are also likely to be more affectively committed to their organizations. Still, it is important to reiterate that neither the raw correlations reported in Table 4.2 nor the initial SEM analysis depicted in Figure 4.1 revealed significant relationships of explanatory style with work attitudes. Clearly, further study of these somewhat counterintuitive and inconsistent, albeit intriguing, results are necessary to assess the validity of the preceding explanation.

Interrelationships among Work Attitudes

As predicted by H_{6a} and H_{6b}, respectively, job satisfaction was positively related to affective commitment, and negatively related to turnover intentions. The anticipated negative relationship between affective commitment and turnover intentions (H_7) was likewise confirmed. Finally, an indirect effect of job satisfaction on turnover intentions, as mediated by affective commitment, emerged as expected (H_{12d}). Importantly, the relationships among these work attitudes are consistent with available theory and prior research (Cropanzano et al., 1993; Martin & Bennett, 1996; Robert et al., 2000; Williams & Hazer, 1986), and hence provide evidence for the construct validity of the measures and the structural validity of the proposed model.

Practical Implications

Research on the consequences of affect at work is still in its infancy. Still, initial findings, including those of this study, suggest that affect has important implications for understanding and managing organizational behavior. For instance, our results confirm that high PA workers tend to be more satisfied with their jobs and committed to the organization, and, hence, less likely to leave the organization. One obvious implication of this finding is that efforts to select employees who are high in PA may be especially worthwhile. Furthermore, the findings that job satisfaction and affective commitment are, in turn, negatively related to turnover intentions, suggest that efforts to increase employees' satisfaction with their jobs and commitment to the organization, should pay benefits to the organization through reduced levels of turnover.

The link between dispositional affectivity and explanatory style established by this study, provided it can be replicated, has important implications for practitioners. Although these variables appear to have a genetic basis, they are nonetheless shaped by environmental factors (Arvey et al., 1989; Seligman, 1990, 2002). This implies that they can be manipulated through changes in environmental contingencies. This is precisely the argument Luthans (2002b) makes in his call for devoting increased scholarly attention to POB states such as optimism and subjective well-being (happiness). Acknowledging that "there is a fine and somewhat controversial and arbitrary line between psychological states and traits" (p. 59), he nevertheless asserts that adequate empirical evidence exists to treat optimism and subjective well-being as state-like constructs that are amenable to change through training and development.

Toward this end, researchers and practicing psychologists have developed attributional training as an intervention strategy designed to alter debilitating attributional patterns (Abramson et al., 1978; Fosterling, 1985). Indeed, several authors view such training as a highly promising means of alleviating individual performance and motivational deficits in organizations (Gardner & Rozell, 2000; Gist & Mitchell, 1992; Martinko & Gardner, 1982, 1987; Seligman, 1990, 2002). Attributional training uses persuasion to alter maladaptive attributional patterns of capable persons who lack confidence in their abilities. By changing attributions from pessimistic to optimistic patterns, efficacy can be raised and affect altered. Thus, workers who perform poorly are discouraged from making internal, stable, and global attributions, and urged to make self-serving attributions instead. They are likewise trained to make internal, stable, and global attributions for success.

Additional guidelines for changing pessimistic explanatory styles to optimistic ones are supplied by Seligman (1990, 2002). He recommends changing the person's beliefs in such situations so that negative motivational and affective consequences can be avoided. For example, when a pessimistic worker discovers that important computer files have been deleted, she should seek other explanations for the adversity, instead of blaming herself. She may speculate that, "Computer viruses are becoming very common. Perhaps my computer is infected with a virus that erased my files." The consequences change drastically with the alteration of beliefs. The user might say, "So it's not my fault the files are gone. But they still might be in the computer somewhere. Maybe I can find a virus detection and data recovery program to help me remove any viruses and retrieve my files." Through this process, pessimists can change helpless behavior to energized, efficacious, and productive behavior (Gardner & Rozell, 2000).

Of course, care must be taken not to cause members to make unrealistic attributions, which falsely raise performance expectations. Taken to the extreme, optimists may deny responsibility for any of their problems and/ or persist with unproductive strategies despite repeated failures or evidence that their approach is not working. Moreover, the negative, albeit weak, relationships between explanatory style and work attitudes suggested by our comparative analysis raises concerns that extreme optimism may actually cause individuals to be less satisfied with their jobs and committed to their organizations. In light of these dangers, calls to consider the comparative advantages of "realistic optimism" over extreme optimism (Peterson, 2000; Schneider, 2001; Seligman, 1990) are especially relevant. To be realistically optimistic means that you take credit for your successes, while recognizing the impact of contributing factors. Where appropriate, you also assume responsibility for your failures, without making debilitating attributions that identify personal, permanent, and pervasive causes for such outcomes.

Our findings also have implications for leaders seeking to develop the positive psychological capabilities of themselves and their associates. Drawing on the intersection of the POB, transformational leadership, and ethical leadership literatures, Luthans and Avolio (2003) have advanced a model of authentic leadership development. They "define authentic leadership in organizations as a process that draws from both positive psychological capacities and a highly developed organizational context, which results in both greater self-awareness and self-regulated positive behaviors on the part of leaders and associates, fostering positive self-development" (p. 243). They argue that authentic leaders are true to themselves, exhibit a seamless correspondence between their espoused values and behavior, and are driven to develop their associates into leaders.

Central to Luthans and Avolio's (2003) conception of authentic leadership are the positive psychological capacities of both leaders and associates, including the POB state of optimism. Indeed, noting its state-like properties, the contagious optimism of great historical leaders, and the emerging research linking optimism to effective leadership (Wunderley, Reddy, & Dember, 1998) and work-related performance (Gillham, 2000; Peterson, 2000; Schneider, 2001; Schulman, 1999; Seligman, 1990, 2002; Wanburg, 1997), Luthans and Avolio argue that optimism is an especially important component in the development of authentic leaders. Moreover, by modeling optimism through their words and deeds, authentic leaders are able to instill optimism in their associates and inspire them toward action. Our preceding discussion suggests another means whereby authentic leaders can facilitate the development of an optimistic outlook among associates: attributional training. That is, to the extent that attributional training is effective in fostering realistic optimism, it represents a powerful tool for authentic leaders seeking to more fully develop their associates.

Future Research Directions

A limitation of the current study is our usage of the intent-to-leave measure, as opposed to actual turnover data. In addition, the cross-sectional design provides a snapshot of the variables at a particular time, but no insight into their causal ordering. Although SEM enables us to compare alternative causal models to identify the model that best "fits" the data, no definitive conclusions about causality can be made. Moreover, because our data were collected from a single source (employees) and using a single data collection method (questionnaire), they are susceptible to potential mono-source and mono-method biases (Podsakoff & Organ, 1986). Future studies with longitudinal designs that measure the study's variables across time periods are needed to overcome these limitations and more fully explicate the causal relationships between affect, explanatory style, and work attitudes. For example, longitudinal designs that measure both trait- and state-affect would make it possible to examine the causal interrelationships of these variables with explanatory style and work attitudes. Will permanent changes in one's explanatory style, such as those sought by attributional training, produce relatively enduring changes in the types of affective states one experiences, as Seligman (1990, 2002) implies? More specifically, will shifts from pessimistic to optimistic styles cause employees to experience higher levels of PA, and lower levels of NA, across situations?

Support for his assertion would reinforce the claims of POB scholars (Luthans, 2002a, 2002b) that optimism and subjective well-being (happiness) constitute state-like qualities that are amenable to training and development. Additionally, research is needed to assess the utility of attributional training for changing maladaptive attributional patterns, and fostering PA at work. If such training were found to be effective, it would provide practitioners with a powerful tool for alleviating motivational and affective deficits suffered by their employees. Viewed from a more positive perspective, such training may provide a practical means whereby the positive psychological strengths of optimism and happiness can be nurtured and developed among employees. Ultimately, it is hoped that this stream of research will create new and positive avenues whereby authentic leaders and concerned organizations can more fully develop and realize the untapped potential of their human resources.

NOTES

1. Support for this research was provided by a Summer Research Grant funded through the Robert M. Hearin Support Foundation; an earlier version of this chapter was presented at the 1999 annual meeting of the Academy of Management, Chicago.
2. The results obtained from the tests of the alternative models are available from the first author upon request.

REFERENCES

Abramson, L. Y., Seligman, M. E. P., & Teasdale, J. (1978). Learned helplessness in humans: Critique and reformulation. *Journal of Applied Psychology, 87,* 32–48.

Allen, N. J., & Meyer, J. P. (1990). The measurement and antecedents of affective, continuance, and normative commitment to the organization. *Journal of Occupational Psychology, 63,* 1–18.

Arbuckle, J. L., & Wothke, W. (1999). *AMOS 4.0 users' guide.* Chicago: SmallWaters Corporation.

Arvey, R. D., Bouchard, T. J., Jr., Segal, N. L., & Abraham, L. M. (1989). Job satisfaction: Environmental and genetic components. *Journal of Applied Psychology, 74,* 187–192.

Berry, D., & Hansen, J. S. (1996). Positive affect, negative affect, and social interaction. *Journal of Personality and Social Psychology, 71,* 796–809.

Brockner, J., & Guare, J. (1983). Improving the performance of low self-esteem individuals: An attributional approach. *Academy of Management Journal, 26,* 642–656.

Buchanan, C. M., & Seligman, M. E. P. (Eds.). (1995). *Explanatory style.* Hillsdale, NJ: Erlbaum.

Byrne, B. M. (2001a). The Maslach inventory: Validating factorial structure and invariance across intermediate, secondary, and university educators. *Multivariate Behavioral Research, 26,* 583–605.

Byrne, B. M. (2001b). *Structural equation modeling with AMOS: Basic concepts, applications, and programming.* Mahwah, NJ: Erlbaum.

Cameron, K. S., Dutton, J. E., & Quinn, R. E. (Eds.). (2003). *Positive organizational scholarship.* San Francisco: Barrett-Kohler.

Campbell, C. R., & Martinko, M. J. (1998). An integrative attributional perspective of empowerment and learned helplessness: A multimethod field study. *Journal of Management, 24,* 173–200.

Carver, C. S., & Scheier, M. F. (1990). Origins and functions of positive and negative affect: A control-process view. *Psychological Review, 97,* 19–35.

Cohen, A. (1993). Organizational commitment and turnover: A meta-analysis. *Academy of Management Journal, 36,* 1140–1157.

Cropanzano, R., James, K., & Konovsky, M. A. (1993). Dispositional affectivity as a predictor of work attitudes and job performance. *Journal of Organizational Behavior, 14,* 595–606.

DePaoli, L., & Sweeny, D. (2000). Further validation of the Positive and Negative Affect Schedule. *Journal of Social Behavior and Personality, 15*(4), 561–568.

Diener, E., & Emmons, R. A. (1985). The independence of positive and negative affect. *Journal of Personality and Social Psychology, 47*, 1105–1117.

Duffy, M. K., & Shaw, J. D. (1998). Positive affectivity and negative outcomes: The role of tenure and job satisfaction. *Journal of Applied Psychology, 83*, 950–959.

Eisner, J. P. (1995). The origins of explanatory style: Trust as a determinant of optimism and pessimism. In G. M. Buchanan & M. E. P. Seligman (Eds.), *Explanatory style* (pp. 49–55). Hillsdale, NJ: Erlbaum.

Fitzgerald, L. F., Drasgow, F., Hulin, C. L., Gelfand, M. J., & Magley, V. J. (1997). Antecedents and consequences of sexual harassment in organizations: A test of an integrated model. *Journal of Applied Psychology, 82*, 578–589.

Fosterling, F. (1985). Attributional retraining: A review. *Psychological Bulletin, 98*, 495–512.

Furnham, A., Sadka, V., & Brewin, C. (1992). The development of an Occupational Attributional Style Questionnaire. *Journal of Organizational Behavior, 13*, 27–39.

Gardner, W. L., & Rozell, E. R. (2000). Computer efficacy: Determinants, consequences, and malleability. *Journal of High Technology Management Research, 11*(1), 109–136.

George, J. M. (1992). The role of personality in organizational life: Issues and evidence. *Journal of Management, 18*, 185–213.

Gillham, J. (2000). *The science of optimism and hope: Research essays in honor of Martin E. P. Seligman.* Philadelphia: Templeton Foundation Press.

Gist, M. E., & Mitchell, T. R. (1992). Self-efficacy: A theoretical analysis of its determinants and malleability. *Academy of Management Review, 12*, 472–485.

Gundlach, M. J., Douglas, S. C., & Martinko, M. J. (2003). The decision to blow the whistle: A social information processing framework. *Academy of Management Review, 28*, 107–123.

Herzberg, F. (1966). *Work and the nature of man.* Cleveland, OH: World.

Ironson, G. H., Smith, P., Brannick, M. T., Gibson, W. M., & Paul, K. B. (1989). Construction of a Job in General scale: A comparison of global, composite, and specific measures. *Journal of Applied Psychology, 74*, 193–200.

Judge, T. (1992). The dispositional perspective in human resource research. *Research in Personnel and Human Resources Management, 10*, 31–72.

Judge, T. A. (1993). Does affective disposition moderate the relationship between job satisfaction and voluntary turnover? *Journal of Applied Psychology, 78*, 395–401.

Judge, T. A., Bono, J. E., Thoresen, C. J., & Patton, G. K. (2003). The job satisfaction-job performance relationship: A qualitative and quantitative review. *Psychological Bulletin, 127*, 376–407.

Kaslow, N. J., Rehm, L. P., Pollack, A. L., & Siegel, A. W. (1988). Attributional style and self-control behavior in depressed and nondepressed children and their parents. *Journal of Abnormal Child Psychology, 16*, 163–175.

Kent, R. L., & Martinko, M. J. (1995). The development and evaluation of a scale to measure organizational attributional style. In M. J. Martinko (Ed.), *Attribution theory: An organizational perspective* (pp. 53–75). Delray Beach, FL: St. Lucie Press.

Levin, K., & Stokes, J. P. (1989). Dispositional approach to job satisfaction: Role of negative affectivity. *Journal of Applied Psychology, 74*, 752–758.

Locke, E. A. (1976). The nature and causes of job satisfaction. In M. D. Dunnette (Ed.), *Handbook of industrial and organizational psychology* (pp. 1297–1349). Chicago: Rand-McNally.

Luthans, F. (2002a). The need for and meaning of positive organizational behavior. *Journal of Organizational Behavior, 23,* 695–706.

Luthans, F. (2002b). Positive organizational behavior: Developing and managing psychological strengths. *Academy of Management Executive, 16,* 57–72.

Luthans, F., & Avolio, B. J. (2003). Authentic leadership development. In K. S. Cameron, J. E. Dutton, & R. E. Quinn (Eds.), *Positive organizational scholarship* (pp. 241–259). San Francisco: Barrett-Koehler.

Martin, C. L., & Bennett, N. (1996). The role of justice judgements in explaining the relationship between job satsifaction and organizational commitment. *Group and Organization Management, 21,* 84–104.

Martinko, M. J. (Ed.). (1995). *Attribution theory: An organizational perspective.* Delray Beach, FL: St. Lucie Press.

Martinko, M. J. (1998). *Organizational Attributional Style Questionnaire: Version 2.* Tallahassee: Florida State University.

Martinko, M. J. (2002). *Thinking like a winner: A guide to high performance leadership.* Tallahassee, FL: Gulf Coast Publishing.

Martinko, M. J., & Gardner, W. L. (1982). Learned helplessness: An alternative explanation for performance deficits. *Academy of Management Review, 7,* 195–204.

Martinko, M. J., & Gardner, W. L. (1987). The leader/member attribution process. *Academy of Management Review, 12,* 235–249.

Martinko, M. J., Gundlach, M. J., & Douglas, S. C. (2002). Toward an integrative theory of counterproductive workplace behavior: A causal reasoning perspective. *International Journal of Selection & Assessment, 10*(1–2), 36–50.

Mathieu, J. E., & Zajac, D. M. (1990). A review and meta-analysis of the antecedents, correlates, and consequences of organizational commitment. *Psychological Bulletin, 108,* 171–194.

Meyer, J. P., & Allen, N. J. (1991). A three-component conceptualization of organizational commitment. *Human Resource Management Review, 1,* 61–98.

Mowday, R. T., Porter, L. W., & Steers, R. M. (1982). *Employee-organization linkages: The psychology of commitment, absenteeism, and turnover.* New York: Academic Press.

Newton, T., & Keenan, T. (1991). Further analyses of the dispositional argument in organizational behavior. *Journal of Applied Psychology, 76,* 781–787.

O'Reilly, C. A., III, Chatman, J., & Caldwell, D. F. (1991). People and organizational culture: A profile comparison approach to assessing person-organization fit. *Academy of Management Journal, 34,* 487–516.

Paradowski, J. H. (2001). Positive affectivity, negative affectivity, and job satisfaction: A meta-analysis. *Dissertation Abstracts International: Section B: The Sciences & Engineering, 61(12–B),* 6749.

Pelled, L. H., & Xin, K. R. (1999). Down and out: An investigation of the relationship between mood and employee withdrawal behavior. *Journal of Management, 25,* 875–895.

Peterson, C. (2000). The future of optimism. *American Psychologist, 55*(1), 44–55.

Peterson, C., & Barrett, L. (1987). Explanatory style and academic performance among university freshmen. *Journal of Personality and Social Psychology, 53,* 603–607.

Peterson, C., Maier, S. F., & Seligman, M. E. P. (1993). *Learned helplessness: A theory for the age of personal control.* New York: Oxford University Press.

Peterson, C., & Seligman, M. E. P. (1984). Causal explanations as a risk factor for depression: Theory and evidence. *Psychological Review, 91,* 347–374.

Peterson, C., Semmel, A., von Baeyer, C., Abramson, L. Y., Metalsky, G. I., & Seligman, M. E. P. (1982). The Attributional Style Questionnaire. *Cognitive Therapy and Research, 6,* 287–300.

Peterson, C., & Villanova, P. (1988). An Expanded Attributional Style Questionnaire. *Journal of Abnormal Psychology, 97,* 87–89.

Podsakoff, P. M., & Organ, D. W. (1986). Self-reports in organizational research: Problems and prospects. *Journal of Management, 12,* 531–544.

Price, J. L. (1977). *The study of turnover.* Ames: Iowa State University Press.

Robert, C., Probst, T. M., Martocchio, J. J., Drasgow, F., & Lawler, J. J. (2000). Empowerment and continuous improvement in the United States, Mexico, Poland, and India: Predicting fit on the basis of the dimensions of power distance and individualism. *Journal of Applied Psychology, 65,* 643–658.

Roethlisberger, F. J., & Dickson, W. J. (1939). *Management and the worker.* Cambridge, MA: Harvard University Press.

Schaubroeck, J., Judge, T. A., & Taylor, I., L. A. (1998). Influence of trait negative affect and situational similarity on correlation and convergence of work attitudes and job stress perceptions across two jobs. *Journal of Management, 24,* 553–576.

Schmukle, S. C., Egloff, B., & Burns, L. R. (2002). The relationship between positive and negative affect in the Positive and Negative Affect Schedule. *Journal of Research in Personality, 36*(5), 463–475.

Schneider, S. L. (2001). In search of realistic optimism: Meaning, knowledge, and warm fuzziness. *American Psychologist, 56*(3), 250–263.

Schulman, P. (1999). Applying learned optimism to increase sales productivity. *Journal of Personal Selling and Sales Management, 19,* 31–37.

Schulman, P., Keith, D., & Seligman, M. E. P. (1993). Is optimism heritable?: A study of twins. *Behavior Research and Therapy, 31,* 569–574.

Seligman, M. E. P. (1990). *Learned optimism.* New York: Pocket Books.

Seligman, M. E. P. (2002). *Authentic happiness: Using the new positive psychology to realize your potential for lasting fulfillment.* New York: Free Press.

Seligman, M. E. P., Abramson, L. Y., Semmel, A., & von Baeyer, C. (1979). Depressive attributional style. *Journal of Abnormal Psychology, 88,* 242–247.

Seligman, M. E. P., & Csikszentmihalyi, M. (2000). Positive psychology. *American Psychologist, 55,* 5–14.

Seligman, M. E. P., & Schulman, P. (1986). Explanatory style as a predictor of productivity and quitting among life insurance agents. *Journal of Personality and Social Psychology, 50,* 832–838.

Shaw, J. D. (1999). Job satisfaction and turnover intentions: The moderating role of positive affect. *Journal of Social Psychology, 139*(2), 242–244.

Shaw, J. D., Duffy, M. K., Abdulla, A., & Singh, R. (2000). The moderating role of positive affectivity: Empirical evidence from bank employees in the United Arab Emirates. *Journal of Management, 26,* 139–154.

Shaw, J. D., Duffy, M. K., Jenkins, G. D., Jr., & Gupta, N. (1999). Positive and negative affect, signal sensitivity, and pay satisfaction. *Journal of Management, 25,* 189–206.

Smith, P. C., Kendall, L. M., & Hulin, C. L. (1969). *The measurement of satisfaction in work and retirement.* Chicago: Rand McNally.

Snyder, C. R., & Lopez, S. J. (Eds.). (2002). *Handbook of positive psychology.* Oxford: Oxford University Press.

Staw, B. M., Bell, N. E., & Clausen, J. A. (1986). The dispositional approach to job attitudes: A lifetime longitudinal test. *Administrative Science Quarterly, 31,* 56–77.

Staw, B. M., & Ross, J. (1985). Stability in the midst of change: A dispositional approach to job attitudes. *Journal of Applied Psychology, 70,* 469–480.

Steel, R. P., & Rentsch, J. R. (1997). The dispositional model of job attitudes revisited: Findings of a 10 year study. *Journal of Applied Psychology, 82,* 873–879.

Steers, R. M., & Mowday, R. T. (1981). Employee turnover and post-decision accommodation. In L. L. Cummings & M. Staw (Eds.), *Research in organizational behavior* (Vol. 3, pp. 235–281). Geenwich, CT: JAI Press.

Tett, R. P., & Meyer, J. P. (1993). Job satisfaction, organizational commitment, turnover intention, and turnover: A path analysis based on meta-analytic findings. *Personnel Psychology, 46,* 259–294.

Walumbwa, F. O., & Lawler, J. J. (2003). Building effective organizations: Transformational leadership, collectivism orientation, work-related attitudes, withdrawal behaviors in three emerging economies. *International Journal of Human Resource Management, 14*(7), 1083–1101.

Wanburg, G. R. (1997). Antecedents and outcomes of coping behavior among unemployed and reemployed individuals. *Journal of Applied Psychology, 82,* 731–744.

Watson, D. (1988a). Intraindividual and interindividual analyses of positive and negative affect: Their relation to health complaints, perceived stress, and daily activities. *Journal of Personality and Social Psychology, 54,* 1020–1030.

Watson, D. (1988b). The viccitudes of mood measurement: Effects of varying descriptors, time frame, and response formats on measures of Positive and Negative Affect. *Journal of Personality and Social Psychology, 55,* 128–141.

Watson, D., Clark, L. A., & Carey, G. (1988). Positive and negative affectivity and their relation to anxiety and depressive disorders. *Journal of Abnormal Psychology, 97,* 346–353.

Watson, D., Clark, L. A., & Tellegen, A. (1988). Development and validation of brief measures of positive and negative affect: The PANAS scales. *Journal of Personality and Social Psychology, 54,* 1063–1070.

Watson, D., & Tellegen, A. (1985). Toward a consenual structure of mood. *Psychological Bulletin, 98,* 219–235.

Whitley, B. E., Jr. (1991). On the psychometric properties of the short form of the Expanded Attributional Style Questionnaire. *Journal of Personality Assessment, 57,* 537–539.

Williams, L. J., & Hazer, J. T. (1986). Antecedents and consequences of satisfaction and commitment in turnover models: A reanalysis using latent variable structural equation methods. *Journal of Applied Psychology, 71,* 219–231.

Wren, D. A. (1994). *The evolution of management thought* (4th ed.). New York: Wiley.
Wunderley, L. J., Reddy, W. B., & Dember, W. N. (1998). Optimism and pessimism in business leaders. *Journal of Applied Social Psychology, 28,* 751–760.

CHAPTER 5

ATTRIBUTION AND BURNOUT

Explicating the Influence of Individual Factors in the Consequences of Work Exhaustion

Jonathon R. B. Halbesleben and M. Ronald Buckley
University of Oklahoma

ABSTRACT

The purpose of this research was to investigate the role of attribution in the experience of burnout. In two samples including approximately 260 working adults, we found support for a model proposed by Moore (2000) that hypothesized that the consequences of burnout depend on the nature of the attributions regarding the causes of burnout. Moreover, we provide evidence that self-serving bias and the actor-observer effect influence the experience of burnout in a manner that can exacerbate future feelings of burnout. These findings incrementally extend our understanding of the interaction between environmental and individual processes in the etiology of burnout.

Burnout is a psychological response to chronic work stress that is characterized by exhaustion (a feeling of depleted mental and physical resources), disengagement (the act of distancing oneself from a job and those associated with the job), and reduced feelings of personal and professional effi-

Attribution Theory in the Organizational Sciences, pages 83–110

cacy (Maslach, 1982; Maslach, Schaufeli, & Leiter, 2001). Burnout has received the attention of researchers and organizational leaders alike, primarily due to its relationship to important organizational/personal outcomes, such as job performance (Halbesleben, 2003; Halbesleben & Bowler, in press; Keijsers, Schaufeli, Le Blanc, Zwerts, & Miranda, 1995; Parker & Kulik, 1995; Wright & Bonett, 1997; Wright & Cropanzano, 1998), withdrawal behaviors (Cherniss, 1992; Parker & Kulik, 1995), job attitudes (Jackson, Turner, & Brief, 1997; Leiter, 1991; Leiter, Clark, & Durup, 1994; Leiter & Maslach, 1988; Wolpin, Burke & Greeglass, 1991), turnover intentions (Geurts, Schaufeli, & de Jonge, 1998; Jackson, Schwab, & Schuler, 1986; Maslach & Jackson, 1984), and negative physiological outcomes (Appels & Mulder, 1991; Melamed, Kushnir, & Shirom, 1992; Shirom, Westman, Shamai, & Carel, 1997).

A great deal of the research concerning burnout has focused on its organizational antecedents, including such variables as role ambiguity and conflict (Burke & Greenglass, 1995; Lee & Ashforth, 1993b; Leiter & Schaufeli, 1996), role overload and underload (Cordes, Dougherty, & Blum, 1997; Leiter, 1991), interpersonal conflict and communication patterns (Becker & Halbesleben, 2003; Leiter & Maslach, 1988), reward systems (Jackson et al., 1986), organizational culture and climate (O'Driscoll & Schubert, 1988), and job structure (Jackson et al., 1986). Far less attention has been expended on the manner in which individual factors contribute to the etiology of burnout.

The emphasis on organizational factors is probably due to the commonly held notion that the environment is a more important factor in the development of burnout than are individual factors (Cherniss, 1993; Hallsten, 1993; Maslach & Schaufeli, 1993). However, while individual factors do not appear to have a direct effect on burnout, researchers have suggested that a variety of individual factors may play a role in moderating the relationship between environmental variables and burnout (cf., Cordes & Dougherty, 1993; Halbesleben & Buckley, in press). These moderating factors include personality (Zellars & Perrewé, 2001; Zellars, Perrewé, & Hochwarter, 2000), social comparison processes (Buunk & Schaufeli, 1993), and the focus of this chapter, attribution (Moore, 2000). Notably, there has been a dearth of studies examining how individual-level variables influence the relationship between burnout and its consequences.

The purpose of this chapter is to advance our understanding of the moderating effect of individual variables in the etiology of burnout by investigating the role that attributions of stress play in burnout. To that end, we review the literature linking attribution to burnout, with a particular focus on Moore's (2000) attribution approach to burnout. We then provide the results of data from two samples designed to investigate the hypothesis that the nature of attributions concerning burnout will influence the relationship between burnout and its consequences in two samples of working adults. We conclude with a discussion of the implications of the findings for both employees and organizations.

ATTRIBUTION AND BURNOUT

The important role of attribution in the stress literature has been underscored by a number of authors. For example, Perrewé and Zellars's (1999) extension of the cognitive appraisal model of stress proposed by Lazarus and Folkman (1984) suggests that one's attribution of stressors determines how they will cope and react to those stressors (see also Zellars, Perrewé, Ferris, & Hochwarter, Chapter 8, this volume). Specifically, they posit that attributions regarding the cause of stress during primary appraisal lead to different emotional reactions, which in turn lead to different coping responses.

Burnout is generally considered a response to work-related stress (Maslach, 1982). As with stress, attribution has long been considered a potentially important aspect of the burnout process (Maslach, 1993). Nonetheless, there has been relatively little theoretical and empirical work concerning the role of attribution in the experience of burnout, particularly the role that attribution plays in the consequences of burnout. The most significant link between attribution theory and burnout was established by Moore (2000) in her theory relating attribution to the consequences of work exhaustion. Her model was the first large-scale explication of attribution as a crucial component in understanding the consequences of burnout. As such, we base our empirical work on the theoretical grounding that she has provided.

The Moore Model

In her model, Moore (2000) isolates exhaustion as the core component of burnout, suggesting that other components of burnout are indeed consequences of the exhaustion and detract from the core psychological experience of those experiencing exhaustion. This approach is justified, in large part, due to two issues. First, while Maslach's (1982, 1993) original three-component conceptualization of burnout has dominated the burnout research (particularly as it relates to measurement of burnout), other conceptualizations have been proposed for the processes underlying burnout (cf., Demerouti, Bakker, Nachreiner, & Schaufeli, 2001; Pines, Aronson, & Kafry, 1981; Shirom, 1989). The alternative conceptualizations of burnout differ in a variety of ways; however, all of them include exhaustion as a primary component of burnout, suggesting that it is indeed a core necessity to the experience of burnout. Second, it is common in the burnout literature to find differential relationships between components of burnout and antecedent or consequent processes; exhaustion appears to be the most consistent in its relationships with other variables (Cordes & Dougherty, 1993; Demerouti et al., 2001; Green, Walkey, & Taylor, 1991; Lee & Ashforth, 1996).

The premise of Moore's (2000) model is that the consequences of burnout are dependent upon an individual's attributions regarding the causes of exhaustion. The development of her model was based on Weiner's (1985, 1986; Weiner, Russell, & Lerman, 1979) work on attributions of success and failure. Moore (2000) argues that work exhaustion is perceived by an employee as an unexpected, novel, and important situation; in effect, it becomes viewed as a failure at work. As a failure, the employee is likely to engage in a search for the causes underlying the exhaustion (Weiner, 1985). When making the attribution regarding the causes of behavior, the employee considers three properties of the situation: causal locus, controllability, and stability (McAuley, Duncan, & Russell, 1982; Moore, 2000; Weiner, 1985).

Causal locus refers to the source, either internal or external to the person, of the cause of a failure or success. In the context of work, an internal attribution of success might be one's perceived ability while an external attribution might relate to perceptions that luck or fate assisted in the success. Controllability suggests that people consider whether they have control over the causes of behavior, particularly when considering the future. They may perceive that they have control (e.g., I have failed this time, but I have the ability to control the outcome for next time) or that others have control (e.g., I failed and have no control over my outcomes). Stability refers to the extent to which the cause is variable or consistent over time. Taken together, these three assessments lead to specific attitudinal, emotional, and behavioral outcomes (Weiner, 1985).

Based on Moore's model and accompanying propositions, we have developed, and later test, a series of hypotheses regarding the relationship between attributions of work exhaustion and the consequences of exhaustion (see also Figure 5.1). We begin with two consequences of exhaustion that Moore argued were independent of attributions, job satisfaction and personal efficacy.

Attribution-Independent Consequences of Exhaustion

Moore (2000) argued that some of the consequences of exhaustion are independent of situational attributions. Specifically, she predicted that outcomes such as lower job satisfaction would be closely related to an employee's

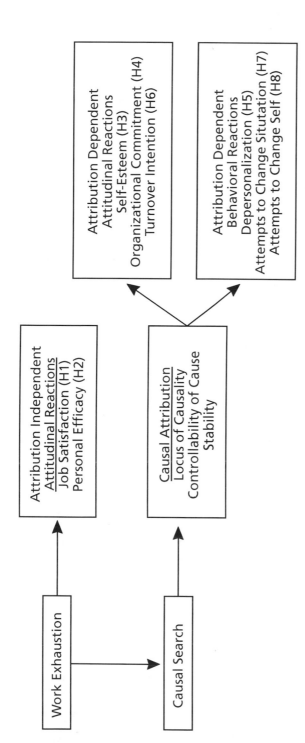

Figure 5.1.   Attributional Model of Burnout (Adapted from Moore, 2000).

experience of exhaustion, but unrelated to an employee's attribution of their exhaustion. This prediction was based on Moore's (1998) previous research and Weiner's (1985) distinction between general and distinct emotions.

Hypothesis 1:   *Work exhaustion will be a significant inverse predictor of job satisfaction.*

It is commonly assumed that work exhaustion is a predictor of the personal efficacy component of burnout (Cordes & Dougherty, 1993). Personal efficacy is considered a component of burnout as it captures the notion that people feel less capable at their jobs when they experience exhaustion. Moore (2000) argued that the reduced personal accomplishment/efficacy dimension of burnout was analogous to a general emotion, and as such, would be predicted by exhaustion but would be unrelated to the attribution of the exhaustion. This is based, in part, on Maslach and Leiter's (1997) contention that workers who experience exhaustion become frustrated by their inability to accomplish their goals.

Hypothesis 2:   *Work exhaustion will be a significant inverse predictor of personal efficacy.*

Attribution-Dependent Consequences of Exhaustion

However, the experience of some consequences of exhaustion will be based on the outcome of the attribution (Moore, 2000). Moore termed these consequences attribution-dependent attitudinal reactions, suggesting that these consequences are wholly dependent on the outcome of the attribution one makes about his or her exhaustion.

Emotional exhaustion has been associated with reductions in self-esteem (Golembiewski & Kim, 1989; McCranie & Bransma, 1988; Rosse, Boss, Johnson, & Crown, 1991). Moore (2000) ties exhaustion and self-esteem through attribution by citing research that suggests that internal attributions of negative events (in this case, exhaustion) are associated with lower self-esteem (Peterson, Maier, & Seligman, 1995; Weiner et al., 1979). She suggests that making an internal attribution about a negative event suggests that the person perceives a deficiency or flaw in oneself; recognition of such a deficiency should lead to reduced feelings of self-esteem.

Hypothesis 3:   *Locus of attributions regarding work exhaustion will be related to self-esteem, where greater internality of attributions will be associated with lower self-esteem.*

A number of studies have reported an inverse relationship between burnout and organizational commitment (cf., Lee & Ashforth, 1996; Leiter & Maslach, 1988). Moore (2000) argued that when employees perceived that their work exhaustion was caused by events that were out of their control and external to them, it would result in lower organizational commitment. She posited that the relationship between such aspects of attribution and commitment was connected with a mismatch between the goals of the organization and the goals of the employee. As a result of this mismatch, the employee feels less commitment to the organization.

Hypothesis 4a: *Locus of attributions regarding work exhaustion will be related to organizational commitment, where greater internality of attributions will be associated with higher organizational commitment.*

Hypothesis 4b: *Controllability of attributions regarding work exhaustion will be related to organizational commitment, where greater perceived controllability by self will be associated with higher organizational commitment.*

A number of burnout researchers have suggested that exhaustion leads to depersonalization (Janssen, Schaufeli, & Houkes, 1999; Koeske & Koeske, 1989; Leiter, 1989; Leiter & Maslach, 1988; Lee & Ashforth, 1993a, 1993b; Maslach & Jackson, 1981; Maslach et al., 2001) and/or disengagement behavior. Depersonalization is manifest when employees begin to psychologically and behaviorally separate themselves from their work. This can be done by simply thinking about work less, mechanistically treating clients as "numbers" or "cases," or more clear behavioral manifestations such as self-isolation while at work. Moore (2000) suggested that this occurs when the employee attributes stability to the cause(s) of their exhaustion. This recognizes that depersonalization occurs when the exhaustion persists over time (Koeske & Koeske, 1989) and as such, an employee might believe the exhaustion will continue in the future. As a result, the employee pulls away from the job and those associated with it (i.e., clients and coworkers) to cope with the anticipated exhaustion.

Hypothesis 5:   *Perceived stability in the attribution of work exhaustion will be positively related to depersonalization.*

Burnout has been associated with an increased intention to leave an organization (Geurts et al., 1998; Jackson et al., 1986). Moore (2000) extended the above logic with regard to organizational commitment and depersonalization to help explain the relationship between attribution of exhaustion and voluntary turnover. Essentially, as an employee realizes his or her goals are a mismatch with the goals of an organization and they are likely to face continued exhaustion, they are more likely to leave the organization. These notions are corroborated by the turnover literature, which has suggested that stable attributions of bad events are likely to increase the likelihood of turnover (Ford, 1985; Seligman & Schulman, 1986), particularly when those stable attributions are perceived to be difficult to change because they are not under the control of the employee (Withey & Cooper, 1989).

Hypothesis 6a: *Perceived stability in the attribution of work exhaustion will be positively related to turnover intention.*

Hypothesis 6b: *Controllability of attributions regarding work exhaustion will be related to turnover intention, where greater perceived controllability by self will be associated with lower turnover intention.*

Moore's (2000) model recognizes that under certain circumstances, employees may choose to stay and change the situation, instead of withdrawing from a situation. Changes in the situation might include attempts to change coworkers and/or supervisors; this may occur through complaining or other persuasion attempts. Moore suggested that the circumstances under which employees choose to attempt situational change are related to the nature of the attribution of exhaustion. Based on work in employee voice (Hirschman, 1970; Hom & Griffeth, 1995; Withey & Cooper, 1989) and learned helplessness (Peterson et al., 1995), she proposed that when an employee attributes his or her exhaustion as external and controllable by others, the employee is more likely to try to change the work situation by voicing concerns to others. As employees see opportunities for change that are outside of themselves (e.g., external and controllable by others), it becomes more appealing to try to change external entities than oneself.

Hypothesis 7a: *Locus of attributions regarding work exhaustion will be related to attempts to change the work situation, where greater internality of attributions will be associated with fewer attempts to change the situation.*

Hypothesis 7b: *Controllability of attributions regarding work exhaustion will be related to attempts to change the work situation, where greater perceived controllability by self will be associated with fewer attempts to change the situation.*

Under other circumstances, the employee will be more likely to try to engage in change related to oneself. These changes might include seeking training or reassessing and changing one's work values and motivation. Based on work by Hom and Griffeth (1995), Moore (2000) proposed that when employees make internal attributions that are associated with perceived controllability by oneself, the employee would attempt to change oneself rather than the situation. In this case, the opportunity for change is seen as within themselves, as such, employees are more likely to engage in self change rather than attempting to change others when such attributions are made. This also fits with work by Brown and Weiner (1984), which suggests that individuals who make internal attributions regarding failures are more likely to engage in personal behavioral changes.

Hypothesis 8a: *Locus of attributions regarding work exhaustion will be related to attempts to change oneself, where greater internality of attributions will be associated with attempts to change oneself.*

Hypothesis 8b: *Controllability of attributions regarding work exhaustion will be related to attempts to change oneself, where greater perceived controllability by self will be associated with attempts to change oneself.*

Social Attribution and Burnout

While Moore's (2000) model clearly explicates the processes underlying self-related attributions of exhaustion, we believe that there are other important social processes related to attribution and burnout that are not included in Moore's model. Specifically, there is a growing interest in the nature of social comparison and attribution processes as they relate to the experience of burnout and its consequences. We believe that the combination of work in social comparison and attribution, specifically the self-serving bias and actor-observer effect, will help to incrementally increase our understanding of the burnout process.

As developed by Jones and Nisbett (1971), the actor-observer effect is the "tendency for actors to attribute their actions to situational requirements, whereas observers tend to attribute the same actions to stable personal dispositions" (p. 2). Nisbett, Caputo, Legant, and Maracek (1973) found evidence of this effect, and suggested that it was caused in part because of a difference in the amount and nature of information that actors possess about observers when compared to them. Gioia and Sims (1985) investigated the actor-observer effect in the context of performance appraisal, and found that subordinates tended to attribute their performance more externally while their managers tended to view the same behavior as internally caused.

The self-serving bias complements the actor-observer effect; it is the tendency for individuals to make external attributions for their failures and internal attributions for their successes. In the context of burnout, the role of the self-serving bias and actor-observer effects is intriguing. Based on the above discussion, it would appear likely that because it is negative, an employee would have an increased likelihood of making an external attribution about their burnout (the self-serving bias). Moreover, employees would be more likely to ascribe internal attributions to their peers' burnout experiences (the actor-observer effect).

Hypothesis 9: *Employees are more likely to make attributions that are external for their own burnout, while making attributions that are internal for burnout experienced by others.*

The intriguing theoretical implication of the combined effects of the self-serving bias and actor-observer effect are the consequent processes. Self-serving bias tends to lead to positive emotional outcomes for those who experience it (Miller & Ross, 1975; Riess, Rosenfeld, Melburg, & Tedeschi, 1981). Additionally, in situations of failure or negative outcomes, the actor-observer effect may also lead to positive emotional outcomes, as the comparison between the actor and observer indicates that the actor's failure is not their fault, while the opposite holds true for the observer.

However, this process does not seem to hold true in burnout. It seems unlikely that an employee who attributes his or her burnout to external causes that are controllable by an external entity will have positive psychological outcomes, particularly given our previous discussion of the outcomes of such attributions in terms of Moore's (2000) model. Moreover, the notion that a coworker's burnout is internally caused by factors the coworker can control (and, by extension, fix if he or she chooses to) seems unlikely to lead to positive emotional outcomes for the employee. Instead, the combination of the self-serving bias and actor-observer effect in burnout may actually result in negative outcomes.

Among those negative outcomes, we expect that the employee would experience greater exhaustion, as they would become further frustrated that they cannot address their own situation to move them toward their goals (Maslach & Leiter, 1997), particularly when it appears that coworkers are better equipped to handle their burnout (Buunk & Schaufeli, 1993; Daniel & Rogers, 1981; Maslach, 1982).

Hypothesis 10:   *Making attributions that are external for their own burnout while making attributions that are internal for burnout experienced by others will be associated with increased levels of exhaustion.*

METHOD

Participants and Procedure

Sample 1. The participants in the first sample were 170 working adults. The sample included 94 males and 76 females with a mean age of 40.19 years of age. The participants had held their current positions for an average of 5.29 years and had been working for their current organization for 8.62 years. A wide variety of industries were represented, including customer service, sales and retail ($n = 29$); health care ($n = 21$); manufacturing/repair ($n = 12$); government and civil service ($n = 9$); insurance ($n = 7$); and education ($n = 6$).

The data were collected with the assistance of introductory management students as part of a research experience assignment. The surveys were completed by the participants and returned to the students. To ensure that the surveys were indeed completed by the working adults, we randomly selected 10 percent of the surveys and directly contacted the participant to verify their participation. Of those contacted, all of the participants verified that they had completed the survey. This method of survey collection has been effectively used by field researchers in organizational settings (cf., Ferris et al., in press; Kolodinsky, Hochwarter, & Ferris, in press).

Sample 2. There were a number of limitations to sample 1 that we sought to address by collecting more data. First, sample 1 was cross-sectional, thus severely limiting any attempt to suggest that attributions caused the outcome variables. Second, we were unable to test the hypotheses (9 and 10) regarding the self-serving bias and actor-observer effect with the data from sample 1. To address these concerns and extend the external validity of our findings, we collected data from an additional sample of working adults.

The participants in sample 2 were 89 working adults. The sample included 47 males and 42 females with a mean age of 35.86 years of age. The participants had held their current positions for an average of 4.11 years and had been working for their current organization for 6.71 years. A wide variety of industries were represented, including manufacturing/repair ($n = 14$), customer service ($n = 11$), sales ($n = 9$), insurance ($n = 8$), and health care ($n = 7$).

The data were again collected with the assistance of introductory management students as part of a research experience assignment. The students collected measures from three working adults at two points during the semester (with approximately one month separating time 1 and time 2 data collection). The burnout inventory and attribution inventory were completed at time 1. The burnout inventory was repeated at time 2. Additionally, the MSQ, OCQ, turnover intention scale, self-esteem scale, and change inventory were completed at time 2. The surveys were completed by the participants and returned to the students.

To ensure that the surveys were indeed completed by the working adults, procedures for verification were employed that were identical to those utilized in sample 1. Again, of those contacted, all of the participants verified that they had completed the survey.

Measures

Attribution. Attribution was assessed using a modified version of the Occupational Attributional Style Questionnaire (OASQ; Furnham, Sadka,

& Brewin, 1992). While we were not specifically concerned with attributional style in this research, the OASQ was used as a model because it provides a stimulus for employees to think about target situations or experiences (in this case, burnout), asks them the cause of the target, and asks them to rate that target on dimensions of attributions derived from Weiner's (1986) attribution theory. We modified the scale such that the target situation asked the participants to think about past experiences of symptoms of exhaustion at work. Overall, the participants were asked to respond to five scenarios, based on the five emotional exhaustion questions of the Maslach Burnout Inventory (see Appendix). They were then asked questions to assess the internality, stability, externality, personal control, and colleague control.

We mean-aggregated the questions about internality and externality (after reverse scoring the externality questions) across scenarios to create a continuous scale of internality and externality of attribution. We mean-aggregated the questions about controllability by self and others (after reverse scoring the controllability by others questions) across scenarios to create a continuous scale of controllability of attribution. Finally, we mean-aggregated questions about stability across scenarios to create one scale of stability of the attribution.

In addition to completing the attribution inventory for themselves, the participants in sample 2 were also asked to complete the attribution inventory considering the burnout of their closest colleague. They completed the same questions as those designed for themselves, with slight wording changes to reflect the symptoms of exhaustion experienced by their closest colleague as the target of the inventory.

Burnout. Burnout was assessed using the Maslach Burnout Inventory—General Survey (MBI-GS; Schaufeli, Leiter, Maslach, & Jackson, 1996). The MBI-GS was used to assess all three dimensions of burnout relevant to the hypotheses: emotional exhaustion, depersonalization, and personal efficacy. The MBI-GS (as opposed to the original MBI) was utilized as it is intended to be a general survey that applies to a wide number of occupations.

Satisfaction. We assessed job satisfaction with the 20-item Minnesota Satisfaction Questionnaire (MSQ; Weiss, Dawis, England, & Lofquist, 1967). The MSQ is a 20-item assessment that measures satisfaction with a variety of aspects of a participant's job (i.e., pay, supervision, coworkers). It is scored on a five-point, Likert-type scale. We took a mean aggregation of the responses on the MSQ to derive a satisfaction score for each participant.

Organizational commitment. We assessed organizational commitment using the Organizational Commitment Questionnaire (OCQ; Mowday, Steers, & Porter, 1979). The OCQ is a 15-item assessment that is scored on a five-point, Likert-type scale. We took a mean aggregation of the responses on the OCQ to derive a commitment score for each participant.

Turnover intention. To assess participants' intention to leave their position, we utilized the turnover intention scale of Mitchell, Holton, Lee, Sablynski, and Erez (2001). The scale is a three-item assessment that is scored on a five-point, Likert-type scale. We took a mean aggregation of the responses on the turnover intention scale to derive a turnover intention score for each participant.

Self-Esteem. The revised Self-Liking and Competence Scale (SLCS-R; Tafarodi & Swann, 2001) is a 16-item instrument designed to measure two dimensions of global self-regard: self-liking (generalized sense of one's worth) and self-competence (generalized sense of one's efficacy). Items are rated using Likert-type scales ranging from 1 (*strongly disagree*) to 5 (*strongly agree*).

Change. To assess attempts to change either the situation or oneself, we developed a change scale. The questions were based on suggestions by Moore (2000) concerning the manner in which one may seek to change the situation or oneself. We derived two questions intended to assess attempts to change the situation through voicing concerns ("In times when you have not been happy with your work situation, how often have you complained to a coworker?" and "In times when you have not been happy with your work situation, how often have you presented suggestions to your supervisor about how to change the situation?") and two questions intended to assess attempts to change oneself ("In times when you have not been happy with your work situation, how often have you sought to change something about yourself [for example, seeking more training]?" and "In times when you have not been happy with your work situation, how often have you asked your supervisor for feedback with the hope of changing yourself?")

RESULTS

The mean, standard deviation, internal reliability estimates, and interscale correlations for each of the variables in the study can be found in Table 5.1. Hypothesis 1 predicted that work exhaustion would be a significant inverse predictor of satisfaction. Using regression analysis, we found support for that hypothesis, where emotional exhaustion scores on the Maslach Burnout Inventory were inversely related to scores on the Minnesota Satisfaction Questionnaire in sample 1 ($\beta = -.28$, $F = 32.96$, $p < .0001$, $R^2 = .17$) and sample 2 ($\beta = -.34$, $F = 29.05$, $p < .0001$, $R^2 = .27$).

Hypothesis 2 predicted that work exhaustion would be a significant inverse predictor of personal efficacy. Using regression analysis, we found support for that hypothesis, where emotional exhaustion scores on the Maslach Burnout Inventory were inversely related to scores on the personal efficacy subscale of the Maslach Burnout Inventory in both sample 1 ($\beta = -.26$, $F = 25.47$, $p < .0001$, $R^2 = .13$) and sample 2 ($\beta = -.28$, $F = 19.69$, $p < .0001$, $R^2 = .18$).

Table 5.1. Sample 1 Scale Means, Standard Deviations, Internal Consistency Estimates, and Interscale Correlations

	Mean	SD	1.	2.	3.	4.	5.	6.	7.	8.	9.	10.	11.	12.
1. Exhaustion	2.62	0.98	.87	-.18*	-.16*	.21**	-.41***	-.36***	-.45***	.45***	.56***	.40***	.32***	-.15
2. Locus of Causality	3.97	1.56		.80	.60***	-.076	.18*	.13	.29***	-.25**	-.16*	-.24**	-.19*	.19*
3. Controllability	4.29	1.56			.83	-.10	.11	.16*	.23**	-.22**	-.094	-.17*	-.21**	.20**
4. Stability	2.91	1.36				.81	-.19*	-.34***	-.27***	.42***	.22**	.14	.18*	-.18*
5. Satisfaction	3.72	0.63					.92	.50***	.76***	-.41***	-.52***	-.51***	-.18*	.13
6. Personal Efficacy	4.16	0.70						.81	.53***	-.51***	-.50***	-.34***	-.19*	.15*
7. Commitment	4.32	0.81							.93	-.48***	-.54***	-.57***	-.29***	.26***
8. Self-esteem	2.04	0.64								.94	.35***	.46***	.31***	-.17*
9. Depersonalization	2.48	0.84									.75	.45***	.18*	-.19*
10. Turnover Intent	2.14	1.31										.91	.24*	-.068
11. Situation Change	2.85	0.99											.74	.019
12. Self Change	2.75	1.00												.74

Note: Internal consistency estimates (Cronbach's alpha) are along the diagonal. * indicates $p < .05$, ** indicates $p < .01$, *** indicates $p < .001$.

Table 5.1. Sample 2 Scale Means, Standard Dev., Internal Consistency Estimates, and Interscale Correlations (Cont.)

	Mean	SD	1.	2.	3.	4.	5.	6.	7.	8.	9.	10.	11.	12.	13.
1. Exhaustion (T1)	2.80	1.07	.92	-.18	-.26*	.38***	-.51***	-.43***	-.60***	-.01	.58***	.56***	.37***	-.36***	.76***
2. Locus (T1)	2.69	1.60		.81	.75***	.05	.020	.050	.30**	-.29**	-.18	-.23*	-.29**	.21*	-.21*
3. Controllability (T1)	3.02	1.61			.77	-.078	.14	.15	.33**	-.32**	-.26*	-.26*	-.28**	.24*	-.27*
4. Stability (T1)	2.81	.64				.78	-.15	-.29**	-.33**	.14	.21*	.22*	.27*	-.18	.38***
5. Satisfaction	3.59	.64					.91	.54***	.83***	.30**	-.64***	-.64***	-.33**	.21*	-.50***
6. Personal Efficacy	3.96	.71						.80	.56***	.073	-.43***	-.42***	-.24*	.27**	-.45***
7. Commitment	4.14	.84							.93	-.096	-.72***	-.74***	-.50***	.37***	-.58***
8. Self-esteem	3.54	.89								.89	-.052	-.040	.28**	-.30**	-.08
9. Depersonalization	2.74	.94									.81	.60***	.35***	-.32**	.49***
10. Turnover Intent	2.21	1.22										.94	.47***	-.11	.52***
11. Situation Change	3.12	1.00											.75	-.26*	.40***
12. Self Change	2.83	1.02												.74	-.39***
13. Exhaustion (T2)	2.91	1.03													.90

Note. Internal consistency estimates (Cronbach's alpha) are along the diagonal. * indicates $p < .05$, ** indicates $p < .01$, *** indicates $p < .001$.

In testing Hypotheses 3–8, we first conducted a MANOVA that included the effects of all three attribution dimensions (locus, controllability, and stability) and all of the dependent variables relevant to those hypotheses. The results of this test indicated the presence of significant effects in both sample 1 ($F = 14.32$, $p < .0001$) and sample 2 ($F = 21.99$, $p < .0001$); as such, we proceeded by testing each of the hypotheses utilizing multiple regression relevant for each hypothesis. In every case, we controlled for the effects of work exhaustion.

Table 5.2 presents the univariate regression results for both samples. Hypothesis 3 predicted that internal attribution would be a significant negative predictor of self-esteem. Using regression analysis, we found support for that hypothesis, where locus of causality scores on the attribution inventory were inversely related to scores on the SLCS-R self-esteem scale in both sample 1 ($\beta = -.10$, $F = 11.08$, $p = .0011$, $R^2 = .062$) and sample 2 ($\beta = -.16$, $F = 8.13$, $p < .0054$, $R^2 = .09$) after controlling for exhaustion.

Table 5.2.   Attribution Dimensions and Consequences of Exhaustion Regression Results

| | Sample 1 | | | Sample 2 | | |
	Locus	Controlla-bility	Stability	Locus	Controlla-bility	Stability
Dependent Variable						
Self-Esteem	–.10**	—	—	–.16**	—	—
Commitment	.12*	.046	—	.07	.12	—
Depersonalization	—	—	.13**	—	—	.11*
Turnover Intention	—	–.13*	.12	—	.14*	–.19*
Situation Change	–.065	–.093	—	–.11	–.08	—
Self Change	–.071	.086	—	–.04	.11	—

Note: Table entries are standardized beta scores. * $p < .05$, ** $p < .01$.

Hypotheses 4a and 4b predicted that external attribution and perceived controllability by others would be a significant inverse predictor of organizational commitment. Using multiple regression analysis, we found partial support for that hypothesis in sample 1; locus of causality scores were significantly associated with the OCQ ($\beta = .12$, $F = 6.66$, $p = .011$) after controlling for exhaustion; however, controllability scores on the attribution inventory were not associated with scores on the Organizational Commitment Questionnaire ($\beta = .046$, $F = 0.92$, $p = .34$, overall $R^2 = .089$). In sample 2, neither locus of causality ($\beta = -.07$, $F = 0.69$, $p = .41$)

nor controllability scores (β = .12, F = 2.28, p = .13, overall R^2 = .11) were associated with OCQ scores.

Hypothesis 5 predicted that stable attributions would be a significant predictor of depersonalization. Using regression analysis, we found support for that hypothesis, where stability scores on the attribution inventory were associated with scores on the depersonalization subscale of the Maslach Burnout Inventory in sample 1 (β = .13, F = 8.37, p = .0043, R^2 = .047) and sample 2 (β = .11, F = 4.13, p = .046, R^2 = .05) after controlling for exhaustion.

Hypotheses 6a and 6b predicted that stable attributions and perceived controllability by others would be a significant predictor of turnover intention. Using multiple regression analysis, we found partial support for that hypothesis in sample 1; controllability scores on the attribution inventory were associated with scores on the turnover intention inventory (β = –.13, F = 4.08, p = .045) after controlling for exhaustion; however, stability scores were not significantly associated with turnover intentions (β = .12, F = 2.46, p = .12, overall R^2 = .04). On the other hand, the data from sample 2 provided support for the hypothesis, with both stability (β = .14, F = 3.88, p = .05) and controllability scores (β = –.19, F = 5.81, p = .018, overall R^2 = .11) associated with turnover intention.

Hypotheses 7a and 7b predicted that external attribution and perceived controllability by others would be a significant inverse predictor of attempts to change the situation. Using multiple regression analysis, we did not find support for that hypothesis with the data from sample 1; neither locus of causality scores (β = –.065, F = 2.16, p = .28) nor controllability scores (β = –.093, F = 2.40, p = .12, overall R^2 = .050) on the attribution inventory were associated with scores on the situational change scale after controlling for exhaustion. These findings were corroborated by the data from sample 2, where neither locus of causality scores (β = –.11, F = 1.46, p = .23) nor controllability scores (β = –.08, F = 0.76, p = .39, overall R^2 = .092) on the attribution inventory were associated with scores on the situational change scale after controlling for exhaustion.

Hypotheses 8a and 8b predicted that internal attribution and perceived controllability by oneself would be a significant inverse predictor of attempts to change oneself. Using multiple regression analysis, we did not find support for that hypothesis in sample 1; neither locus of causality scores (β = –.071, F = 1.39, p = .24) nor controllability scores (β = .086, F = 2.00, p = .15, overall R^2 = .049) on the attribution inventory were associated with scores on the self change scale after controlling for exhaustion. The data from sample 2 also did not support the hypothesis, as neither locus of causality scores (β = –.04, F = 0.21, p = .65) nor controllability scores (β = .11, F = 1.35, p = .25, overall R^2 = .059) on the attribution inventory were associated with scores on the self change scale after controlling for exhaustion.

Hypothesis 9 predicted that employees would be more likely to make attributions that are external and characterized by controllability by others for their own burnout and attributions that are internal and characterized by controllability by the observer for their colleague's burnout. To evaluate this we tested the difference between the sample 2 participant's assessment of the locus of causality for their own symptoms of exhaustion and the participant's assessment of the locus of causality for the symptoms of exhaustion of his or her peers. We found support for the hypothesis, where participants' attribution scores represented a trend toward more externality than they attributed to their colleagues, $t(1, 85) = -4.73$, $p < .001$.

Finally, hypothesis 10 predicted that the self–other difference found above would predict time 2 exhaustion scores. Using regression analysis, we found support for that hypothesis, where the difference between attribution styles of participants and their perceptions of peers were associated with time 2 scores on the exhaustion subscale of the Maslach Burnout Inventory ($\beta = .14$, $F = 5.29$, $p = .024$, $R^2 = .059$).

GENERAL DISCUSSION

The results from the two samples provide some mixed findings regarding the outcomes of burnout when considering the nature of attributions. While many of the findings supported Moore's (2000) predictions, most of which were consistent across the two samples, there are a number of exceptions. We found that exhaustion was related to lower job satisfaction and lower personal efficacy. On the other hand, self-esteem, depersonalization, and turnover intention were associated with specific patterns of attribution in response to exhaustion. Finally, we found evidence for a pattern whereby participants were more likely to make internal and self-controllable attributions for their own burnout, while making external and outwardly controllable attribution for their peers. This pattern was associated with increased exhaustion over a one-month span.

Perhaps most notable was the lack of association between predicted attribution components and attempts to change either the situation or oneself. There are a number of potential explanations for the results. First, the change scale was the only scale that we developed. As such, there is limited evidence to support its validity, particularly in terms of entirely capturing the construct of change (such restricted content validity may be reflected in the relatively low reliabilities of these scales). Future research that refines the measurement of change attempts, including incorporating observations from those external to the participant, would be particularly valuable.

Second, from a theoretical perspective, there is uncertainty about the direct effect of attribution on change efforts. While our hypotheses follow Moore's (2000) model and propositions, her model also included a link from attitudinal outcomes of attributions (e.g., commitment, reduced personal efficacy, and reduced self-esteem) to behavioral outcomes that included change efforts. However, she did not specify the nature of that link. As such, the prediction of change efforts is likely improved when considering those variables, but further theoretical development is needed to guide research in that area.

One potential manner to address these concerns, both theoretically and methodologically, would be to consider the role of coping from the stress literature. For example, rather than considering attempts to change oneself or attempts to change the situation, problem-focused and emotion-focused coping might be more useful constructs. Problem-focused coping behaviors include seeking information about what needs to be done and then taking action as a result (Folkman, Lazarus, Dunkel-Schetter, DeLongis, & Gruen, 1986). On the other hand, emotion-focused coping is manifest in distancing oneself from work (which is conceptually similar to depersonalization), avoiding stressors, and emphasizing the positive aspects of situations (Folkman & Lazarus, 1985). As has been noted by Zellars et al. (Chapter 8, this volume; see also Perrewé & Zellars, 1999), these coping strategies are related to attributions about one's attributions and emotions; considering these types of coping behaviors within Moore's (2000) attributional framework of burnout may provide a more clear tie to stress theory and research than attempts to change self versus situation. This is particularly relevant as these coping behaviors have been linked in the literature to other constructs relevant to the study of burnout, such as work–family conflict (cf., Rotondo, Carlson, & Kincaid, 2003).

A surprising finding came with respect to organizational commitment. With one exception, the hypotheses regarding the relationship between attribution and organizational commitment were not supported. There are a number of potential explanations for these findings. Questions have been raised regarding the construct validity of the OCQ; as such, measurement limitations may be masking the actual relationships between attribution of exhaustion and commitment. However, it should be noted that Moore's (2000) proposal specifically postulated a relationship between locus and controllability of attributions and *affective* commitment, the component of commitment for which the OCQ is most suited for measurement. Perhaps this is a hypothesized relationship that requires revision.

Implications for Burnout Research and Theory

While much of this study supports Moore's (2000) model via empirical evidence for the relationships between attribution and the consequences of exhaustion, there are a number of important considerations in terms of theoretical development that need to be addressed. Among those is the specific relationship between the dimensions of attribution. In her presentation of the model, Moore is unclear about these relationships; the manner in which her propositions are proposed (and thus, our hypotheses) suggests that they are additive in nature. In response, we have tested the hypotheses in this fashion. However, one could argue that the dimensions are related multiplicatively (interactively) , and as such, tests of interaction effects may be more appropriate. To test this idea, we entered the interactions between attribution dimensions into the regression equations, following the proposed tests of Cohen and Cohen (1983) for moderation. In all cases, the interaction terms were not significant beyond the individual effects of the attribution dimensions, limiting empirical support for the argument that the dimensions of attribution interact to predict the outcomes of exhaustion.[1]

Further confusion comes from the specific relationship between exhaustion, the dimensions of attribution of exhaustion, and the consequences of exhaustion. The model and propositions proposed by Moore (2000) suggest a mediated relationship (e.g., exhaustion leads to causal dimensions, which then directly lead to consequences); however, it is unclear whether the relationship is mediated or moderated (e.g., exhaustion leads to causal search, but the actual outcome of the causal search in terms of dimensions moderates the relationship between exhaustion and its consequences). More work, both theoretical and empirical, is necessary to work through this critical issue in the model.

Another area of development that is needed with regard to Moore's (2000) model is integration with other theories of stress and burnout. As an example, the model can be explained in terms of the conservation of resources (COR) model that has been put forth by Hobfoll (1988, 1989). The COR model suggests that burnout occurs in response to either the loss of resources or the threat of resource loss. As a result of that loss, employees tend to take steps to protect their resources. As such, many of the consequences of stress and burnout are attempts to conserve resources. For example, employees may tend to withdraw from their work in response to burnout (disengagement) as a mechanism for conserving interpersonal resources. More integration is needed with the COR model, as it holds potential for addressing some of the limitations of Moore's model. Specifically, the COR model might help explain why we did not find links from attribution and change efforts of the participants. One might argue that to attempt change might be a risky investment of resources, and as such, would not be likely.

An important contribution of this study is the consideration of the role that self-serving biases and the actor-observer effect play on the attribution process and the outcomes of burnout. This represents the first attempt to empirically test these attribution errors, and their resultant effect, in the context of burnout. Research in this topic is important because it integrates attribution and social comparison research to help expand our understanding of the burnout process. Burnout researchers have begun to recognize the important role of social comparison in the experience of burnout (see Buunk & Schaufeli, 1993, for a review); this research integrates that line of thinking with the emergent view that attribution influences the experience of burnout.

Implications for Practice

From a practical perspective, this study suggests that understanding the nature of the attributions employees make regarding their burnout can have an important influence on the consequences of burnout. While organizations should strive to provide organizational environments that reduce the likelihood of burnout (what some have termed "healthy organizations"; Cox, Kuk, & Leiter, 1993), managers must also be cognizant of the important role that employees' attributions can play in the burnout process. More specifically, managers need to recognize that employees' responses to their stress and burnout are important indicators of how they will deal with burnout. By recognizing the cognitive responses of employees to their stress, efforts at dealing with that stress can begin, particularly if the cognitive response is an inaccurate representation of the environment.

Managers can also better address burnout by understanding the courses taken to reduce its consequences. For example, in situations where employees are attempting to change themselves or the situation, a manager might recognize this as a response to exhaustion, and specifically, attributions regarding the exhaustion. By influencing both the causes of the exhaustion and working with the employee to address the perceptions of its causes, managers may help to address potential problems with turnover, lowered organizational commitment, reduced personal efficacy, and depersonalization behaviors. Moreover, they may be able to better utilize change efforts such that employees will see a greater benefit.

Limitations

There are a number of limitations to the design of the present research that should be noted. First, in asking participants to recall situations where they have experienced symptoms of exhaustion (see Appendix), there is the possibility that they have never actually experienced the symptoms, and as such, are not in a position to reflect on the causes of the symptoms. This argument can be further extended when considering sample 2, where participants were asked to consider the causes of exhaustion symptoms of their coworkers. Given the reported prevalence of stress and burnout in organizations, including characterizations such as "epidemic" and "pandemic" (Golembiewski, Boudreau, Munzenrider, & Luo, 1996; Maslach & Leiter, 1997; Schaufeli, in press; Schaufeli & Enzmann, 1998), we suspect it is unlikely that employees have not experienced these symptoms. However, we recognize that individuals who are at the extreme levels of burnout may have different reactions than their relatively less burned-out counterparts. As such, our study provides a rather conservative test, by including a majority of participants who would not be considered extremely burned out. Research that more specifically investigates the attribution patterns of significantly burned-out employees would be valuable in addressing this concern; however, we note the methodological difficulties in undertaking such research, as those experiencing extreme burnout are typically no longer employed, and may be experiencing other mental health difficulties, such as extreme depression, that would preclude them from more typical organizational research (Schaufeli, in press).

There are a number of other methodological and statistical limitations that we recognize as potential confounds in interpreting our results. By collecting information from the same participants using self-report measures, we run the risk of common method bias influencing the relationships between the variables under study (Podsakoff, MacKenzie, Lee, & Podsakoff, 2003; Podsakoff & Organ, 1986; Williams, Cote, & Buckley, 1989). While collecting data over two points of time in sample 2 helps reduce some of these concerns, they are not entirely remedied and may serve to influence the results.

Finally, our sampling technique has generated a nonrandom convenience sample of working adults that leads to two concerns. The sample may not be representative of the population of working adults, which may limit the extent to which our findings generalize. Moreover, this leads to potential limitations in terms of normality of data distributions that serve as an assumption for regression. We do note, however, that by sampling from a wide variety of occupational groups and levels within multiple organizations, we believe that we have captured more of the variability in worker experiences than may be found in typical, single-organization studies.

CONCLUSION

This research provides support for Moore's (2000) general notion that the outcomes of burnout are not necessarily universally experienced; instead, they are dependent on the nature of the attributions one makes regarding the burnout. This is significant in that it demonstrates that the interaction between environmental and individual factors may be a significant cause of burnout. While we certainly agree with the notion that environmental factors are primary (as has been aptly noted by others, "you have to be in the fire to get burned"), research continues to suggest that individual factors certainly contribute to the burnout process. This study offers empirical support for the notion that attribution is a key individual factor that influences both the experiences of and the consequences of burnout.

NOTE

1.   These analyses are available by request from the first author.

ACKNOWLEDGMENTS

The authors would like to acknowledge Jenn Becker, Constance Campbell, Bernard Weiner, Terence Mitchell, Mark Martinko, and three anonymous reviewers for their comments on previous drafts of this chapter and Tosha Morris and Brandy Dodson for their assistance with data entry.

APPENDIX

Exhaustion Attribution Scale Scenarios

Think about the last time...

- You felt emotionally exhausted by your work
- You felt used up at the end of your workday
- You felt tired when you got up in the morning and had to face another day on the job
- Working all day really felt like a strain to you
- You felt burned out from your work

Note. For the questions used in sample 2 to test social attribution hypotheses, the questions were reworded such that "You" was replaced with "Your closest coworker."

REFERENCES

Appels, A., & Mulder, P. (1991). Burnout as a risk factor for coronary heart disease. *Behavioral Medicine, 17*, 53–59.

Becker, J. A. H., & Halbesleben, J. R. B. (2003). *Defensive communication and burnout in the workplace: The mediating role of leader-member exchange.* Paper presented at the annual meeting of the National Communication Association, Miami Beach, FL.

Brown, J., & Weiner, B. (1984). Affective consequences of ability versus effort ascriptions: Controversies, resolutions, and quandaries. *Journal of Educational Psychology, 76,* 146–158.

Burke, R. J., & Greenglass, E. R. (1995). A longitudinal study of psychological burnout in teachers. *Human Relations, 48,* 187–202.

Buunk, B. P., & Shaufeli, W. B. (1993). Burnout: Perspective from social comparison theory. In W. B. Schaufeli, C. Maslach, & T. Marek (Eds.), *Professional burnout: Recent developments in theory and research* (pp. 53–66). Washington, DC: Taylor & Francis.

Cherniss, C. (1992). Long-term consequences of burnout: An exploratory study. *Journal of Organizational Behavior, 13,* 1–11.

Cherniss, C. (1993). Role of professional self-efficacy in the etiology and amelioration of burnout. In W. B. Schaufeli, C. Maslach, & T. Marek (Eds.), *Professional burnout: Recent developments in theory and research* (pp. 135–149). Washington, DC: Taylor & Francis.

Cohen, J., & Cohen, P. (1983). *Applied multiple regression/correlation analyses for the behavior sciences* (2nd ed). Hillsdale, NJ: Erlbaum.

Cordes, C. L., & Dougherty, T. W. (1993). A review and integration of research on job burnout. *Academy of Management Review, 18,* 621–656.

Cordes, C. L., Dougherty, T. W., & Blum, M. (1997). Patterns of burnout among managers and professionals: A comparison of models. *.Journal of Organizational Behavior, 18,* 685–701.

Cox, T., Kuk, G., & Leiter, M. P. (1993). Burnout, health, work stress, and organizational healthiness. In W. B. Schaufeli, C. Maslach, & T. Marek (Eds.), *Professional burnout: Recent developments in theory and research* (pp. 177–193). Washington, DC: Taylor & Francis.

Daniel, S., & Rogers, M. L. (1981). Burn-out and the pastorate: A critical review with implications for pastors. *Journal of Psychology and Theology, 9,* 232–249.

Demerouti, E., Bakker, A. B., Nachreiner, F. & Schaufeli, W. B. (2001). The job demands-resources model of burnout. *Journal of Applied Psychology, 86,* 499–512.

Ferris, G. R., Treadway, D. C., Kolodinsky, R. W., Hochwarter, W. A., Kacmar, C. J., & Douglas, C. (in press). Development and validation of the political skill inventory. *Journal of Management.*

Folkman, S., & Lazarus, R. S. (1985). If it changes it must be a process: Study of emotion and coping during three stages of a college examination. *Journal of Personality and Social Psychology, 48,* 150–170.

Folkman, S., Lazarus, R. S., Dunkel-Schetter, C., DeLongis, A., & Gruen, R. J. (1986). Dynamics of a stressful encounter: Cognitive appraisal, coping, and encounter outcomes. *Journal of Personality and Social Psychology, 50,* 992–1003.

Ford, J. D. (1985). The effects of causal attributions on decision makers' responses to performance downturns. *Academy of Management Review, 10,* 770–786.

Furnham, A., Sadka, V., & Brewin, C. R. (1992). The development of an occupational attributional style questionnaire. *Journal of Organizational Behavior, 13,* 27–39.

Geurts, S., Schaufeli, W., & de Jonge, J. (1998). Burnout and intention to leave among mental health-care professionals: A social psychological approach. *Journal of Social and Clinical Psychology, 17,* 341–362.

Gioia, D. A., & Sims, H. P. (1985). Self-serving bias and actor-observer differences in organizations: An empirical analysis. *Journal of Applied Social Psychology, 15,* 547–563.

Golembiewski, R. T., Boudreau, R. A., Munzenrider, R. F., & Luo, H. (1996). *Global burnout: A world-wide pandemic explored by the phase model.* Greenwich, CT: JAI Press.

Golembiewski, R. T., & Kim, B. S. (1989). Self-esteem and phases of burnout. *Organizational Development Journal, 7,* 51–58.

Green, D. E., Walkey, F. H., & Taylor, A. J. W. (1991). The three-factor structure of the Maslach Burnout Inventory. *Journal of Social Behavior and Personality, 6,* 453–472.

Halbesleben, J. R. B. (2003). *Burnout and job performance: The mediating role of motivation.* Paper presented at the annual meeting of the Academy of Management, Seattle, WA.

Halbesleben, J. R. B., & Bowler, W. M. (in press). Organizational citizenship behaviors and burnout. In D. L. Turnipseed (Ed.), *A handbook on organizational citizenship behavior: A review of 'good soldier' activity in organizations.* Hauppauge, NY: Nova Science.

Halbesleben, J. R. B., & Buckley, M. R. (in press). Burnout in organizational life. *Journal of Management.*

Hallsten, L. (1993). Burning out: A framework. In W. B. Schaufeli, C. Maslach, & T. Marek (Eds.), *Professional burnout: Recent developments in theory and research* (pp. 95–113). Washington, DC: Taylor & Francis.

Hirschman, A. O. (1970). *Exit, voice, and loyalty: Reponses to decline in firms, organizations, and states.* Cambridge, MA: Harvard University Press.

Hobfoll, S. E. (1988). *The ecology of stress.* New York: Hemisphere.

Hobfoll, S. E. (1989). Conservation of resources: A new attempt at conceptualizing stress. *American Psychologist, 44,* 513–524.

Hom, P. W., & Griffeth, R. W. (1995). *Employee turnover.* Cincinnati, OH: South-Western.

Jackson, S. E., Schwab, R. L., & Schuler, R. S. (1986). Toward an understanding of the burnout phenomenon. *Journal of Applied Psychology, 71,* 630–640.

Jackson, S. E., Turner, J. A., & Brief, A. P. (1987). Correlates of burnout among public service lawyers. *Journal of Occupational Behavior, 8,* 339–349.

Janssen, P. P. M., Schaufeli, W. B., & Houkes, I. (1999). Work-related and individual determinants of the three burnout dimensions. *Work and Stress, 13,* 74–86.

Jones, E. E., & Nisbett, R. E. (1971). *The actor and the observer: Divergent perceptions of the causes of behavior.* Morristown, NJ: General Learning Press.

Keijsers, G. J., Schaufeli, W. B., Le Blanc, P. M., Zwerts, C., & Miranda, D. R. (1995). Performance and burnout in intensive care units. *Work and Stress, 9,* 513–527.

Koeske, G. F., & Koeske, R. D. (1989). Construct validity of the MBI: A critical review and reconceptualization. *Journal of Applied Behavioral Science, 25,* 131–144.

Kolodinsky, R. W., Hochwarter, W. A., & Ferris, G. R. (in press). Nonlinearity in the relationship between political skill and work outcomes: Convergent evidence from three studies. *Journal of Vocational Behavior.*

Lazarus, R. S., & Folkman, S. (1984). *Stress, appraisal, and coping.* New York: Springer.

Lee, R. T., & Ashforth, B. E. (1993a). A further examination of managerial burnout: Toward an integrated model. *Journal of Organizational Behavior, 14,* 3–20.

Lee, R. T., & Ashforth, B. E. (1993b). A longitudinal study of burnout among supervisors and managers: Comparisons between the Leiter and Maslach (1988) and Golembiewski et al. (1986) models. *Organizational Behavior and Human Decision Processes, 54,* 369–398.

Lee, R. T., & Ashforth, B. E. (1996). A meta-analytic examination of the correlates of the three dimensions of job burnout. *Journal of Applied Psychology, 81,* 123–133.

Leiter, M. P. (1989). Conceptual implications of two models of burnout. *Group and Organizational Studies, 14,* 15–22.

Leiter, M. P. (1991). Coping patterns as predictors of burnout: The function of control and escapist coping. *Journal of Occupational Behavior, 12,* 123–144.

Leiter, M. P., Clark, D., & Durup, J. (1994). Distinct models of burnout and commitment among men and women in the military. *Journal of Applied Behavioral Science, 30,* 63–82.

Leiter, M. P., & Maslach, C. (1988). The impact of interpersonal environment on burnout and organizational commitment. *Journal of Occupational Behavior, 8,* 297–308.

Leiter, M. P., & Schaufeli, W. B. (1996). Consistency of the burnout construct across occupations. *Anxiety, Stress, and Coping, 9,* 229–243.

Maslach, C. (1982). *Burnout: The cost of caring.* Englewood Cliffs, CA: Prentice-Hall.

Maslach, C. (1993). Burnout: A multidimensional perspective. In W. B. Schaufeli, C. Maslach, & T. Marek (Eds.), *Professional burnout: Recent developments in theory and research* (pp. 19–32). Washington, DC: Taylor & Francis.

Maslach, C., & Jackson, S. E. (1981). The measurement of experienced burnout. *Journal of Occupational Psychology, 2,* 99–115.

Maslach, C., & Jackson, S. E. (1984). Patterns of burnout among a national sample of public contact workers. *Journal of Health and Human Resource Administration, 7,* 189–212.

Maslach, C., & Leiter, M. P. (1997). *The truth about burnout.* San Francisco: Jossey-Bass.

Maslach, C., & Schaufeli, W. B. (1993). Historical and conceptual development of burnout. In W. B. Schaufeli, C. Maslach, & T. Marek (Eds.), *Professional burnout: Recent developments in theory and research* (pp. 1–18). Washington, DC: Taylor & Francis.

Maslach, C., Schaufeli, W. B., & Leiter, M. P. (2001). Job burnout. *Annual Review of Psychology, 52,* 397–422.

McAuley, E., Duncan, T. E., & Russell, D. W. 1992. Measuring causal attributions: The revised Causal Attribution Scale (CDSII). *Personality and Social Psychology Bulletin, 18,* 566–573.

McCranie, E. W., & Bransma, J. M. (1988). Personality antecedents of burnout among middle-aged physicians. *Behavioral Medicine, 36,* 889–910.

Melamed, S., Kushnir, T., & Shirom, A. (1992). Burnout and risk factors for cardiovascular disease. *Behavioral Medicine, 18,* 53–61.

Miller, D. T., & Ross, M. (1975). Self-serving biases in the attribution of causality: Fact or fiction? *Psychological Bulletin, 82,* 213–225.

Mitchell, T. R., Holton, B. C., Lee, T. W., Soblynski, L. J., & Erez, M. (2001). Why people stay: Using job embeddedness to predict voluntary turnover. *Academy of Management Journal, 44,* 1102–1122.

Moore, J. E. (1998). An empirical test of the relationship of causal attribution to work exhaustion consequences. In M. A. Rahim, R. T. Golembiewski, & C. C. Lundberg (Eds.), *Current topics in management* (Vol. 3, pp. 49–67). Greenwich, CT: JAI Press.

Moore, J. E. (2000). Why is this happening? A causal attribution approach to work exhaustion consequences. *Academy of Management Review, 25,* 335–349.

Mowday, R. T., Steers, R. M., Porter, L. W. (1979). The measurement of organizational commitment. *Journal of Vocational Behavior, 14,* 224–247.

Nisbett, R. E., Caputo, C., Legant, P., & Marecek, J. (1973). Behavior as seen by the actor and as seen by the observer. *Journal of Personality and Social Psychology, 27,* 154–164.

O'Driscoll, M. P., & Schubert, T. (1988). Organizational climate and burnout in a New Zealand social service agency. *Work and Stress, 2,* 199–204.

Parker, P. A., & Kulik, J. A. (1995). Burnout, self- and supervisor-rater job performance and absenteeism among nurses. *Journal of Behavioral Medicine, 18,* 581–599.

Perrewé, P. L., & Zellars, K. L. (1999). An examinations of attributions and emotions in the transactional approach to the organizational stress process. *Journal of Organizational Behavior, 20,* 739–752.

Peterson, C., Maier, S. F., & Seligman, M. E. P. (1995). *Learned helplessness: A theory for the age of personal control.* New York: Oxford University Press.

Pines, A., Aronson, E., & Kafry, D. (1981). *Burnout: From tedium to personal growth.* New York: Free Press.

Podsakoff, P. M., MacKenzie, S. B., Lee, J. & Podsakoff, N. P. (2003). Common method biases in behavioral research: A critical review of the literature and recommended remedies. *Journal of Applied Psychology, 88,* 879–903.

Podsakoff, P. M., & Organ, D. W. (1986). Self-reports in organizational research: Problems and prospects. *Journal of Management, 12,* 531–544.

Riess, M., Rosenfeld, P., Melburg, V., & Tedeschi, J. R. (1981). Self-serving attributions: Biased private perceptions and distorted public descriptions. *Journal of Personality and Social Psychology, 41,* 224–231.

Rosse, J. G., Boss, R. W., Johnson, A. E., & Crown, D. F. (1991). Conceptualizing the role of self-esteem in burnout. *Group and Organization Studies, 16,* 428–451.

Rotondo, D. M., Carlson, D. S., & Kincaid, J. F. (2003). Coping with multiple dimensions of work-family conflict. *Personnel Review, 32,* 275–296.

Schaufeli, W. B. (in press). Past performance and future perspectives on burnout research. *South African Journal for Industrial and Organisational Psychology.*

Schaufeli, W. B., & Enzmann, D. (1998). *The burnout companion to study and practice.* London: Taylor & Francis.

Schaufeli, W. B., Leiter, M. P., Maslach, C., & Jackson, S. E. (1996). Maslach Burnout Inventory-General Survey (MBI-GS). In C. Maslach, S. E. Jackson, & M. P. Leiter, *MBI Manual* (3rd ed.). Palo Alto, CA: Consulting Psychologists Press.

Seligman, M. E. P., & Schulman, P. (1986). Explanatory style as a predictor of productivity and quitting among life insurance sales agents. *Journal of Personality and Social Psychology, 50,* 832–838.

Shirom, A. (1989). Burnout in work organizations. In C. L. Cooper & I. Robertson (Eds.), *International review of industrial and organizational psychology* (pp. 25–48). New York: Wiley.

Shirom, A., Westman, M., Shamai, O., & Carel, R. S. (1997). Effects of work overload and burnout on cholesterol and triglycerides levels: The moderating effects of emotional reactivity among male and female employees. *Journal of Occupational Health Psychology, 2,* 275–288.

Tafarodi, R. W., & Swann, W. B., Jr. (2001). Two-dimensional self-esteem: Theory and measurement. *Personality and Individual Differences, 31,* 653–673.

Weiner, B. (1985). An attributional theory of achievement motivation and emotion. *Psychological Review, 92,* 548–573.

Weiner, B. (1986). *An attributional theory of motivation and emotion.* New York: Springer-Verlag.

Weiner, B., Russell, D., & Lerman, D. (1979). The cognition-emotion process in achievement-related contexts. *Journal of Personality and Social Personality, 37,* 1211–1220.

Weiss, D. J., Dawis, R. V., England, G. W., & Lofquist, L. H. (1967). *Manual for the Minnesota Satisfaction Questionnaire.* Minneapolis: University of Minnesota Industrial Relations Center.

Williams, L. J., Cote, J. A., & Buckley, M. R. (1989). Lack of method variance in self-reported affect and perceptions at work: Reality or artifact? *Journal of Applied Psychology, 74,* 462–468.

Withey, M. J., & Cooper, W. H. (1989). Predicting exit, voice, loyalty, and neglect. *Administrative Science Quarterly, 34,* 521–539.

Wolpin, J., Burke, R. J., & Greenglass, E. R. (1991). Is job satisfaction an antecedent or a consequence of psychological burnout? *Human Relations, 44,* 193–209.

Wright, T. A., & Bonett, D. G. (1997). The contribution of burnout to work performance. *Journal of Organizational Behavior, 18,* 491–499.

Wright, T. A., & Cropanzano, R. (1998). Emotional exhaustion as a predictor of job performance and voluntary turnover. *Journal of Applied Psychology, 83,* 486–493.

Zellars, K. L., & Perrewé, P. L. (2001). Affective personality and the content of emotional social support: Coping in organizations. *Journal of Applied Psychology, 86,* 459–467.

Zellars, K. L., Perrewé, P. L., & Hochwarter, W. A. (2000). Burnout in health care: The role of the five factors of personality. *Journal of Applied Social Psychology, 30,* 1570–1598.

CHAPTER 6

CORE SELF-EVALUATIONS, ASPIRATIONS, SUCCESS, AND PERSISTENCE

An Attributional Model

Timothy A. Judge and John D. Kammeyer-Mueller
University of Florida

ABSTRACT

The study of attributions and personality are two of the most well-developed areas in all of psychology, but there are only limited efforts to integrate these areas due to the division between experimental and correlational psychology. The literature on attributions has also been divided into affective and cognitive camps. To achieve rapprochement between these areas, a model is developed that proposes that attributions are affected by stable core self-evaluations, and that these attributions, in turn, affect more proximal self-evaluations. The resultant model provides an opportunity to restore the concept of process to a central role in personality research and understand how stable individual differences might affect attributions for specific events.

Understanding causal relationships is fundamental to the way that human beings make sense of and attempt to adapt to their worlds, even though for

Attribution Theory in the Organizational Sciences, pages 111–132

the most part, we are unable to observe causes directly (Einhorn & Hogarth, 1986). David Hume (1748/1963) argued that cause-and-effect relationships do not exist in nature *a priori*, although it is the nature of human consciousness to determine the causes of events. He wrote:

> Our conviction of the truth of a fact rests on feeling, memory, and the reasonings founded on the causal connection, i.e. on the relation of cause and effect. The knowledge of this relation is not attained by reasonings a priori, but arises entirely from experience.... Hence there is no knowledge and no metaphysics beyond experience.

Because experiences that inform causal inferences were influenced by feelings and sentiments, Hume's philosophy provides a tacit role for individual differences in causal attributions. While Hume's assertions regarding the centrality of experience have been controversial, alternative philosophical schools go even further in emphasizing the uniquely constructive role of the individual in forming an understanding of cause (e.g., Kant, 1781/1998; Leibniz, 1765/1982).

Yet, within the realm of psychology, the study of attributions often has ignored individual differences and individual subjectivity. As noted by Allport (1955) and Cronbach (1957), there is a long history of differentiating the field of psychology into individual differences research that primarily concentrates on stable differences between people (e.g., personality psychology) and intraindividual differences research that concentrates on how situations affect behavior (e.g., social psychology). Traditionally, attributions have been the domain of social psychologists (Weiner, 1990). Indeed, some of the more renowned social psychologists played pivotal roles in the development of attribution theory,[1] including Heider (1958) and Kelley (1967, 1973). Unfortunately, because of the differences in methods and focus for personality and social psychology, there is a history of either suspicion or overt challenge between these "two disciplines of psychology," and very little integration of research between fields. Forgas, Bower, and Moylan (1990) noted that "attribution researchers have paid relatively little attention to individual differences and the personal states and characteristics of judges, such as their emotional states" (p. 809).

As noted by Weiner and Graham (1999), attribution theory need not be under the exclusive purview of social psychologists interested in cognitive reactions to situations. Specifically, Weiner (1990) notes that attributions may both affect, and be affected by, personality processes. For example, Weiner and Graham (1999, p. 605) noted, "Answers to a question such as 'Why have I failed?' surely can affect self-esteem.... In addition, self-esteem is likely to influence the answer to that question." Moreover, as noted by

Martinko (1995), only limited, relatively recent research has studied attribution theory in the organizational sciences.

The purpose of this chapter is to provide a model that integrates contemporary concepts in personality and motivation with attribution theory concepts. Our goal in doing this is to augment the existing literature, which has only described the cognitive framework for making attributions, with affective and experiential components (Forgas et al., 1990). In the next section of this chapter, we introduce and discuss the meaning and relevance of the concepts that appear in the model. In subsequent sections, we present hypotheses linking these concepts in the context of an integrated model. Finally, in the last section of this chapter we discuss how future research could productively test the model and the relationships embedded therein. Our goal is to shed better light on the *intra*personal processes that may be substantially informed by attribution theory, but which have been relatively ignored in research (Martinko, 1995).

ATTRIBUTION THEORY

Attributions lie at the core of human reasoning because they are the principal means by which individuals make sense of their and others' behavior (Einhorn & Hogarth, 1986). It is not surprising, then, that attributions were among the earliest phenomena considered by social psychologists. Heider (1944) believed that all individuals were, in some sense, naïve psychologists who formed beliefs or hypotheses about the motives of themselves and others with whom they interacted, and acted on the basis of these beliefs. Heider (1944, 1958) proposed rules by which responsibility for an action are likely to be attributed to a person. The result of applying these rules range from a fully internal attribution (the person is wholly responsible) to a completely external attribution (the situation is solely responsible). Person factors include ability, motivation, and personality. Situation factors include luck, influential others, and other elements of the environment.

Kelley (1967) extended and formalized Heider's theory by providing specific hypotheses regarding factors that affect the formation of attributions. Specifically, Kelley hypothesized that attributions hinged on three types of information:

1. *Consensus*—how does one's behavior compare to that of one's peers? Consensus is high when one acts similarly to one's peers; it is low when it is different.

2. *Distinctiveness*—how does one's behavior compare to one's behavior in other situations? Distinctiveness is high when one's behavior in

one situation is different from one's behavior in other situations; distinctiveness is low when one's behavior on a task is similar to behavior in other situations.

3. *Consistency*—how consistent is one's behavior on a task over time? Consistency is high when one's behavior is similar over time; consistency is low when one's behavior varies considerably over time.

Kelley argued that individuals ascribe behavior to internal causes when consensus is low, distinctiveness is low, and consistency is high. Individuals will make external attributions when consensus is high, distinctiveness is high, and consistency is low.

Weiner (1980) extended this attribution model to achievement-oriented behavior. In Weiner's model, after an individual performs a task, he or she seeks to judge whether it was successful or unsuccessful and determines what factors might have caused that success or failure. Weiner hypothesized that attributions would hinge on three factors: locus of control (internal vs. external), stability (whether causes change over time), and controllability (whether the causes can be changed by the person). The results of this attributional process produce changes in an individual's self-concept, which then produce changes in behavior. Evidence indicates that people's attributions of their current performance foreshadow their expectations concerning future performances (e.g., Forsyth & McMillan, 1981). Martinko and Thomson (1998) provided an integration of Kelley's and Weiner's theories, based on the idea that both models describe the same fundamental attribution process. This model begins to move toward suggesting an important role for self-construal in determining behavior and attributions. A more comprehensive integration of the self-concept requires a more detailed understanding of personality, however. This is the topic to which we turn in the next section of the chapter.

PERSONALOGICAL STATES AND TRAITS

The concept of personality is among the most central and diffuse in all of psychology. While nearly any topic in the study of human beings could potentially be studied under the topic of personality, for the most part personality researchers have been concerned with the sense of an integrated self and personal consistency over time and across situations. Allport (1937) described personality as the "dynamic organization within the individual of those psychophysical systems that determine his unique adjustments to his environment" (p. 48). For Allport (1955), the dynamics of the internal process were what made personality psychology distinctive, so much so that he proposed that intra-individual growth and change should be central to the study of personality. Similar psychologies of intra-individual change were investigated by other seminal writers of personality theory (e.g., Erikson, 1959; Murray, 1938).

However, a review of the recent literature shows that intra-individual variation is not currently a major component of personality research in the organizational sciences. Instead, researchers have expended far more time and effort in an attempt to define the taxonomy of important traits on which individuals differ reliably across situations. This theme has been so central that to many individuals, personality is considered the study of traits. This research has amply demonstrated that individuals differ on aggregate measures of personality, and that these differences are consistent over time (e.g., Costa & McCrae, 1997; Roberts & DelVecchio, 2000; Watson & Walker, 1996). On the environmental side, psychodynamic researchers propose that early life experiences have an especially powerful imprinting on an individual, which is unlikely to change over time (Westen, 1990). Moreover, research shows that there is a genetic component to personality traits, further enhancing the argument for stability and consistency in behavior (Loehlin, 1992; Loehlin, McCrae, Costa, & John, 1998).

However, the existence of stable, genetically inherited traits does not necessarily mean that personality is immutable. As noted in Allport's definition of personality provided earlier, there is an important dynamic aspect to personality. Even if personality does show impressive rank order stability, there is considerable change over time in each person's alignment and relationship between traits—a well-adjusted adult of 40 probably behaves quite differently than a similarly well-adjusted 20-year-old even though they might have quite similar trait scores (Allport, 1955). Life cycle researchers focus not only on how events affect personality, but how interpretations of events build on one another across the life span (McAdams, 1990). An increasing number of researchers in the area of personality psychology have turned to an interactional approach, wherein there is an interplay between person factors and situational factors (e.g., Magnusson, 1990). The dynamic between the two has been captured by work in the area of anxiety (Spielberger, Gorsuch, Lushene, Vagg, & Jacobs, 1983) and more general measures of negative affect (Watson, Clark, &Tellegen, 1988), which explicitly focus on the relationship between higher-order trait measures and day-to-day states. Steyer, Ferring, and Schmitt (1992) also demonstrate that for both anxiety and coping behaviors, models that use both state and trait properties are consistently superior than those that focus on only one. This research suggests that incorporating some element of the effect of the environment from social psychology might greatly enhance our understanding of personality.

Another key area of interest for personality theorists is the factors that influence one's self-construal, particularly in the form of self-appraisals of self-esteem and self-efficacy. This is the most direct area for attribution theory's contribution. Cognitive theories of personality tend to focus on how individuals construct a view of reality through selective attention to various aspects of their environment and understanding causal linkages (Mischel, 1990). Social cognitive theory similarly takes a person's self-observations of behavior as a starting point for subsequent self-construals, which in turn will influence self-efficacy and behavior (Bandura, 1991). More affect-based models of self-appraisal note that individuals with more negative self-images tend to concentrate on personal shortcomings as explanations for poor performance (Di Paula & Campbell, 2002; Dodgson & Wood, 1998).

Given the apparent linkages between personality and attribution theory, we further propose that researchers may already have begun to explore one of the most central personality traits for the study of attributions. In the next section, we review core self-evaluations, which we propose is the single most important constellation of personality traits for attribution theory, and which we also believe is the component of personality that is most likely to be affected by attributions.

CORE SELF-EVALUATIONS

At the nexus of the relationship between a person's construal of their environment and their exercise of personal agency lies a person's construal of themselves. While diverse research has treated self-evaluations along dimensions of efficacy, esteem, locus of control, and neurosis, Judge, Locke, and Durham noted in 1997 there is a common core to all these dimensions that can be termed *core self-evaluations*. According to Judge and colleagues (1997), core self-evaluations are defined as fundamental premises that individuals hold about themselves and their functioning in the world (Judge & Larsen, 2001). In the 6 years since the publication of the first paper on the topic, more than 20 articles (e.g., Erez & Judge, 2001), chapters (e.g., Bono & Judge, 2003), and dissertations (e.g., Best, 2003; Erez, 1997; Rode, 2002) have been conducted, addressing issues ranging from the construct validity of the trait to its role in explaining and predicting job satisfaction and job performance. Although Judge and colleagues originally linked core self-evaluations to job satisfaction (Judge, Locke, Durham, & Kluger, 1998), more recent research has linked the concept to life satisfaction (Heller, Judge, & Watson, 2002), job performance (Judge & Bono, 2001), and motivation (Erez & Judge, 2001; Judge, Erez, & Bono, 1998). In this research, the core self-evaluations concept emerges as a consistently valid predictor of both affective and objective work outcomes.

In their initial formulation of the core self-evaluations concept, Judge and colleagues (1997) searched the literature for traits that met three criteria: *self-evaluative* (core traits should involve self-appraisal as opposed to description of oneself or others), *fundamentality* (core traits should be fundamental as opposed to surface traits; Cattell, 1965), and *scope* (core traits should be wide in scope, or cardinal traits; Allport, 1961). Using these criteria, Judge and colleagues (1997, 1998) proposed core self-evaluations as a higher-order concept comprised of four more specific lower order traits: (1) *self-esteem*—the basic appraisal people make of themselves and the overall value that people place on themselves; (2) *generalized self-efficacy*—individuals' estimate of their fundamental ability to cope with life's exigencies, to perform, and to be successful; (3) *locus of control*—the degree to which individuals believe that they control events in their lives; and (4) *neuroticism* (or its converse, *emotional stability*)—one of the "Big Five" personality dimensions that represents the tendency to exhibit poor emotional adjustment and experience negative affects such as fear, hostility, and depression (Goldberg, 1990).

As might be expected given their conceptual similarities, empirically, the traits are strongly interrelated. In a meta-analysis, Judge, Erez, Bono, and Thoresen (2002) revealed the following correlations between the traits:

- Self-esteem–locus of control, $\hat{\rho}$ =.52.
- Self-esteem–emotional stability, $\hat{\rho}$ =.64.
- Self-esteem–generalized self-efficacy, $\hat{\rho}$ =.85.
- Locus of control–emotional stability, $\hat{\rho}$ =.40.
- Locus of control–generalized self-efficacy, $\hat{\rho}$ =.56.
- Emotional stability–generalized self-efficacy, $\hat{\rho}$ =.62.

The average (absolute) correlation among the traits is .60. Furthermore, the four traits appear to indicate a single common factor, and the measures load strongly on the common factor (e.g., average loading = .80; Erez & Judge, 2001), and this general factor appears to be more useful in predicting various criteria (such as job satisfaction and job performance) than the specific variance attributioned to the subtraits (Judge et al., 2002).

Based on the emerging body of evidence suggesting that core self-evaluations is the best representation of a person's self-construal, we believe this construct offers a unique opportunity to build on the existing literature on self-concept and attribution theory. One of the key advantages of the core self-evaluations construct is that it brings together several dimensions of a person's self-concept that might be relevant to performance. It captures the common variance to variables considered relatively cognitive (i.e., self-efficacy and locus of control) along with variables of a more affective or motivational nature (i.e., self-esteem and emotional stability). In addition to its value as a representation of several traits underlying the self-concept, the core self-evaluations model provides a meaningful link to several related literatures that have previously examined the role of self-perception and performance from the distinct perspectives offered by self-esteem, self-efficacy, locus of control, and emotional stability. As such, research on core self-evaluations promises to help unify several distinct methods of conceptualizing the effects of attributions on self-concept and performance, thereby adding to the literature on each of the related sub-constructs. Because attribution theories have been linked to the core traits of self-esteem (Weiner, 1987) and locus of control (Rotter, 1966), it is apropos to explore the link between core self-evaluations and attributions. In the next section of this chapter, we discuss attribution theory and its relevance to core self-evaluations research.

MODEL AND HYPOTHESES

Although each of the Kelley (1967) and Weiner (1980) attribution concepts could be studied, here we focus on internal attributions because this stream of research is most germane to current personality research. Internal attributions are common to the attribution theories, including Kelley's theory, Weiner's theory (as locus of causality), and learned helplessness theory (Abramson, 1979; Seligman & Schulman, 1986). Even more specific attributional concepts—the actor-observer bias (Jones & Davis, 1965) and the self-serving bias (Heider, 1958)—focus on attributing causes to the person (internal attribution) or the situation (external attribution). Additionally, of the core concepts in the Kelley, Weiner, and Seligman models, the internality/externality dimension is most related to existing personality constructs. Weiner (1990) noted the study of individual differences in causal attributions began with Rotter's (1966) incorporation of locus of control into the personality literature as an enduring disposition. Another core trait, self-esteem, also has been an important individual difference in research on attribution theory, especially as it relates to threats to self-esteem (Feick & Rhodewalt, 1997) and the hedonic bias (Campbell & Sedikides, 1999).

Figure 6.1 contains the hypothesized model that links trait core self-evaluations, task aspirations and success, internal versus external attributions, state core self-evaluations, and task persistence/withdrawal. The focus of the model is on core self-evaluations (trait and state) and attributions that

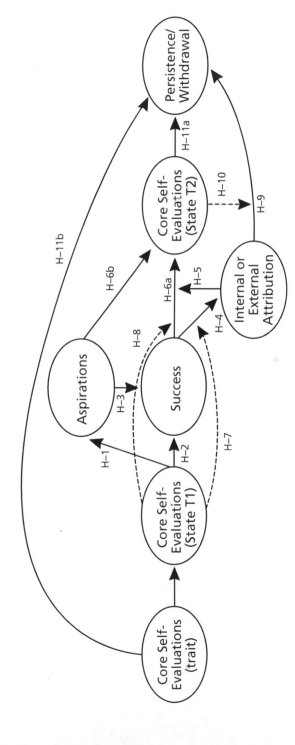

Figure 6.1.   Core self-evaluations: Attributional Model.

may serve as a central mechanism explaining how core self-evaluations are linked to motivation and performance. Below we explicate each of the hypothesized linkages in the model. While several of the concepts presented in this model have been described previously (e.g., Donovan & Williams, 2003; Eden, 1988; Locke & Latham, 1990; Thomas & Mathieu, 1994), our goal is to integrate these findings under a single theoretical model that recognizes similarities in the assembled literatures using attributional processes as the focal construct. Although many links in the model have been previously supported through empirical research using related concepts, we contend that there is additional utility to examining whether the combined core self-evaluations construct provides a superior explanatory framework relative to the traits examined in isolation.

Moving from left to right in the model, trait core self-evaluations is linked to both aspirations (H-1) and success (H-2). Although our choice of wording is intentionally broad here,[2] one may think of aspirations as goals or desired end-states and success as task performance. The link between core self-evaluations and goal-setting is clear. Despite impressive support for goal-setting theory (see Locke & Latham, 1990), goals will only motivate people to the degree they are accepted (Locke, 1968). Hollenbeck and Klein (1987) argue that commitment to goals is a function of the expectancy of goal attainment (people will not be committed to goals they think they cannot achieve) and the valence of goal attainment (people will only strive to achieve goals they find attractive). It follows that individuals with positive core self-evaluations should set higher goals because they have greater expectancy of attaining their goals (Earley & Lituchy, 1991; Hollenbeck & Brief, 1987; Phillips & Gully, 1997; Thomas & Mathieu, 1994). Moreover, in a field study, Erez and Judge (2001) found the core self-evaluations trait was positively related to self-set goals (r = .42, $p < .01$) as well as goal commitment (r = .59, $p < .01$).

Hypothesis 1: *Core self-evaluations will be positively related to aspirations such that individuals with a positive self-regard will set higher performance goals than those with a less positive self-regard.*

In terms of core self-evaluations and success, given the link to motivation, it is not surprising that core self-evaluations would be linked to performance. Judge and Bono (2001) linked the four core self-evaluation traits to job performance in a meta-analysis of 105 correlations. The weakest correlation was emotional stability (.19); the strongest correlation was self-esteem (.26). Across the four traits, the average correlation was .23, which is the same as the validity of conscientiousness in predicting job performance (Barrick & Mount, 1991).

Hypothesis 2:   *Core self-evaluations will be positively related to success such that individuals with a positive self-regard will perform better than those with a less positive self-regard.*

The link between aspirations, in the form of goals, and performance is well documented. According to Locke (1997), goals lead to performance because they direct attention and action, arouse effort, and facilitate persistence. As Locke, Shaw, Saari, and Latham (1980) comment, "The beneficial effect of goal setting on task performance is one of the most robust and replicable findings in the psychological literature" (p. 145). Indeed, several meta-analyses support the relationship between both self-set and assigned goals as predictors of job performance (Mento, Steel, & Karren, 1987; Tubbs, 1986). Wright's (1990) review revealed corrected correlations of r = .36 between assigned goals and performance and r = .28 between self-set goals and performance.

Hypothesis 3:   *Aspirations will be positively related to success such that individuals who set higher goals will perform better than those who set lower goals.*

In proffering what has come to be known as the self-serving bias, Heider (1958) argued that individuals attribute their successes to internal causes (one's ability or motivation) and their failures to external causes (chance, task difficulty, influence of others). A meta-analysis of the self-serving bias literature revealed that, across all studies, the average effect size was $\bar{d}$ = .47 (Campbell & Sedikides, 1999), which would translate into $\bar{r}$ = .23. The authors conclude, "Individuals do make internal attributions for their successes and external (person or situation) attributions for their failures" (p. 35). This finding has also received experimental support from research showing self-enhancement motives dominate individuals' self-evaluative motives compared to self-verification or self-assessment (Sedikides, 1993). Thus, as shown in Figure 6.1, and as articulated in H-4 below, we hypothesize that individuals will attribute successful performance to themselves and unsuccessful performance to others or to the situation.[3]

Hypothesis 4:   *Success will be positively related to internal (vs. external) attributions such that individuals who perform well will be more likely to attribute their performance to internal factors (e.g., disposition) than to external factors (e.g., situation).*

Although the revised learned helplessness theory hypothesizes that the *general* tendency to make internal attributions should be positively related to self-esteem and negatively related to depression because it reflects a more optimistic explanatory style (Abramson, Seligman, & Teasdale, 1978), in reacting to specific events, the functionality of attributions depends on the positivity of the event (Aspinwall & Leaf, 2002). Success leads to positive emotions and heightened self-worth when attributions are internal because one is assuming credit for success (Weiner, 1990). Conversely, the same attribution leads to negative emotions and undermines self-worth when the event is negative because one is blaming oneself for failure (Weiner, 1990). In discussing locus of causality and self-esteem, Weiner and Graham (1999) note, "Success outcomes that are ascribed to the self (e.g., personality, ability, effort) result in greater self-esteem and pride" (p. 615). There is also an affective component to these responses—internal attributions to success often lead to states like happiness and relaxation (Weiner, Russell, & Lerman, 1978). Research in clinical psychology suggests that individuals who make internal attributions for positive life events experience fewer depressive symptoms, such as reducing negative affective states and increasing personal efficacy (Needles & Abramson, 1990). Thus, internal attributions should foster positive core self-evaluations when the event is positive (performance above expectations) but undermine positive core self-evaluations when the event is negative (performance below expectations).

Hypothesis 5:   *The effect of success on state core self-evaluations will be moderated by internal or external attributions, such that the effect of internal attributions on state core self-evaluations will be positive when performance is above the aspiration level but negative when performance is below the aspiration level.*

William James (1890) theorized that self-esteem was the ratio of one's successes to one's pretensions or aspirations. Identity theory would predict that one's self-regard is particularly affected by success or failure when the domain is salient to one's identity and when the self-concept is measured in a role-specific manner (Ervin & Stryker, 2001). When the task is important (central to one's identity), the self-esteem threat produced by failure is heightened. Rosenberg and colleagues (1995) show, for example, that high school grades have a stronger relationship with academic self-esteem than global self-esteem. Affect is more centrally involved in processing of information, which is closely related to success or failure on tasks that are peripheral to one's self-concept (Sedikides, 1995). By extension, the relationship between specific self-evaluations and success or failure on a specific task should be more affectively toned than would the relationship between overall self-concept and specific successes or failures. Thus, state core self-evaluations in both the cognitive and affective components should be positively related to success when the self-evaluations and success are assessed in a commensurate manner. Moreover, controlling for this success, one's aspirations will bear a negative relationship to state core self-evaluations because, following James's hypothesis, higher aspirations holding success constant will lead to lower self-esteem. As he noted, "To give up pretensions is as blessed a relief as to get them gratified" (James, 1890).

Hypothesis 6:   *State core self-evaluations will be affected by (a) success and (b) aspirations such that individuals will have the most positive self-regard when their performance exceeds their goals.*

As was noted previously, the self-serving bias—the tendency to make internal attributions for success and external attributions for failure—has received general support in the literature (Campbell & Sedikides, 1999). Yet, there are also potential moderators of this effect. One possible moderator is one's core self-concept. Specifically, it seems likely that one of the ways that positive individuals maintain their favorable self-image is to discount failures (to causes outside themselves) and take credit for successes. Indeed, evidence suggests that individuals with high self-esteem respond to negative feedback through an accentuated self-serving bias (Baumeister, Heatherton, & Tice, 1993). For example, one study revealed that individuals who attributed their recent unemployment to external factors had higher levels of self-esteem than those who gave internal attributions (Winefield, Tiggemann, & Winefield, 1992). Campbell and Sedikides's (1999) review revealed that the self-serving bias was strong for high self-esteem individuals (d = 1.05) but essentially nil for low self-esteem individuals (d = −.07). There is also evidence that individuals in positive moods are more likely to attribute their successes to internal or stable causes (Curren & Harich, 1993; Forgas et al., 1990), suggesting that the emotional stability component of core self-evaluations will act similarly to the self-esteem component by increasing the internality of attributions of success. Thus, success should be more likely to translate into internal attributions for individuals with a positive self-concept than those with a less positive self-concept.

Hypothesis 7:   *Core self-evaluations will moderate the effect of success on internal (vs. external) attributions such that performance is more likely to translate into internal attributions for those with a positive self-regard than for those with a negative self-regard.*

Self-verification theory (Swann, Stein-Seroussi, & Giesler, 1992) maintains that individuals are motivated to preserve their self-image such that individuals with high self-esteem are motivated to find evidence of success, whereas individuals with low self-esteem are motivated to find evidence of failure. Empirical evidence supports self-verification when the self-views are implicated in the feedback individuals receive (Bosson & Swann, 1999).[4] The notion of congruency in information processing is also presented from research on mood and self-concept, which shows that individuals tend to seek out mood-consistent information so that positive individuals will be more sensitive to positive information while negative individuals will be more sensitive to negative information (Mayer, Gaschke, Braverman, & Evans, 1992; Sedikides, 1992). Individuals in negative moods are also more likely to hold themselves to unrealistic standards for feeling that they have performed adequately, leading them to be more likely to see a given level of performance as a failure (Cervone, Kopp, Schaumann, & Scott, 1994). Given the relationship between core self-evaluations and emotional stability, this provides a broader base of support for a relationship than would the self-esteem findings alone. Therefore, individuals with positive trait core self-evaluations could maintain or enhance their state core self-evaluations by making internal attributions for their success and external attributions for their failures. In such a way, positive people reinforce their positive self-image by taking credit for successes and escaping responsibility for failures, and negative people reinforce their negative self-image by blaming themselves for failures and crediting others (or the environment) for successes.

Hypothesis 8:   *The effect of success on subsequent (time 2) state core self-evaluations will be moderated by prior (time 1) state core self-evaluations, such that performance will more strongly affect individuals' subsequent state core self-evaluations for those who are initially high on state core self-evaluations.*

The relationship between attributional style and persistence is complex. On the one hand, learned helplessness theory predicts that those who attribute adverse events to internal, stable, and global causes will persist less in the face of failure, pain, or negative feedback (Abramson et al., 1979). Thus, all else equal, that the tendency to make internal attributions is positively related to persistence (Seligman & Schulman, 1986). However, a general explanatory style is not the same as attributions in reactions to specific episodes. It is hard to imagine that internal attributions when one is clearly not at fault, or persistence in the face of no hope of success, are adaptive. Weiner's (1980) theory predicts that internal attributions in the face of failure will lead to depression and withdrawal. Thus, with respect to specific situations (e.g., repeated performance of the same task), making internal attributions for failure would seem to support withdrawal from the task, whereas making internal attributions for success would support persistence.

Hypothesis 9:   *For successful performance, internal attributions will be related to persistence (H-9a). For unsuccessful performance, internal attributions will be related to withdrawal (H-9b).*

The process described in H-9 could be expected to be accentuated by core self-evaluations. Specifically, the functional process of persisting after success, when one makes an internal attribution, may be stronger for those with a positive self-concept. One of the reasons positive people might better translate internal attributions into persistence in light of successful performance is that they are reward-sensitive. Specifically, positive people may more strongly approach positive outcomes (like the prospect of future successful performance) than less positive people (Snyder, 2002). In support, Erez and Judge (2001) found that the core trait was positively related to goal valence (the attractiveness of goal attainment). Another process supporting this relationship is the expectancy of (future) success. Positive people should be more likely to believe that their capabilities will translate into future success, as evidenced by research showing links between the individual core traits and expectancy motivation (Hollenbeck & Brief, 1987).[5]

Hypothesis 10:   *Core self-evaluations will moderate the effect of internal attributions on persistence on a task such that the effect of internal attributions on persistence following success will be stronger for those with a positive self-regard than those with less positive self-regard.*

A final consideration in the model is the relationship between states and traits as antecedents of behavior. The difficulties in matching attitudes to behavior encountered in social psychology serves as a useful guide in this regard. After nearly half a century of research that produced largely equivocal results on the question of whether attitudes predict behavior, Ajzen and Fishbein (1977) proposed a correspondence model to explain when relationships should be strongest. Their review of the published literature showed that behaviors involve (a) a specific action, (b) performed toward a target, (c) in a context, and (d) at a time or occasion, and that the relationship between attitudes and behaviors will be enhanced to the extent that these elements are overlapping. Similarly, self-evaluations can be expected to be most predictive of behavior to the extent that they refer to an evaluation of the self in regard to a proximal behavior. "Global self-esteem is shown to relate to overall psychological well-being, role-specific self-esteem more directly to behavior" (Ervin & Stryker, 2001, p. 36). Indeed, Rosenberg, Schooler, Schoenbach, and Rosenberg (1995) found that specific self-esteem better predicted behavior to which it was matched (i.e., academic self-esteem and grades) than did global self-esteem.

Hypothesis 11:   *State core self-evaluations (H-11a) will have a more proximal effect on task persistence/withdrawal than will trait core self-evaluations (H-11b).*

DISCUSSION

In summary, the proposed model of attributions and personality provides several linkages that have not previously been explored. While Hypotheses 1–4 have largely been demonstrated in the past, the full model of the relationship between core self-evaluations and performance has not been investigated systematically. The model centers around a proposition made at the beginning of this chapter—namely, that evaluations of the self will exert a powerful influence on how causal forces are interpreted, and that in turn, the interpretation of causation will affect self-evaluations. It is through these reciprocal effects that personality traits, which may begin from inauspicious beginnings as isolated events and inert genetic code, come to be solidified over time through a process of self-reinforcement (Li, 2003).

The model makes several contributions to our understanding of attributions and core self-evaluations. Regarding attributions, although the phenomenological nature of inferences of causality has long been recognized, there has been little systematic work to understand how individual differences might affect how attributions are made. The present model locates both trait- and state-level self-evaluations as a personality dimension that might have especially strong implications for how internality or externality of cause is inferred. Beyond simply suggesting that there will be these effects, the model hypothesizes a sequence for performance as a mediator of the relationship between individual differences and attributions of causality. In addition, the research on attributions has not integrated cognitive and affective processes in the past.

Regarding core self-evaluations, although there is research demonstrating a consistent stability in these evaluations, there is far less research showing how self-evaluations are formed. Most research has implicitly treated core self-evaluations as unchanging properties of the self, without acknowledging how the traits of self-esteem, self-efficacy, and neuroticism have been studied as time varying and contextual. The notion of a state core self-evaluation that combines transitory versions of the core traits of self-esteem, self-efficacy, neuroticism, and locus of control is a potentially important addition to the literature. Although state core self-evaluations are our primary objects of study that might be affected by experiences, the possibility that self-evaluations are the outcome of experience is also a major refinement of core self-evaluations theory that may help extend the research into an understanding of personal agency in general.

As this study is an attempt to bridge a gap between the worlds of the two empirical disciplines of psychology (Cronbach, 1957), empirical evidence for this model will need to come from a variety of sources. One clear implication of the model for research is that meaningful tests of the model will require measurement of global and specific traits. It is our hope that extending the research in this way will facilitate an understanding of how the critical concept of agency is inferred from the environment.

NOTES

1. As noted by Kelley and Michela (1980), there is no single attribution theory; it may be more accurate to refer to attribution *theories* (Martinko & Thomson, 1998).

2. We chose "aspirations" and "success" rather than "goals" and "performance" in the model because the latter terms, while useful, may be overly narrow. Aspirations is the generalized "upward desire for excellence" (Oxford University Press, 2003), which is a broader concept than goal-setting. Core self-evaluations may be linked to aspirations beyond goals in that goals are task specific (Locke, 1997), whereas aspirations may generalize beyond a single task. Moreover, core self-evaluations may be linked for various definitions of success beyond job/task performance, including career success such as earnings or status (Kammeyer-Mueller, Judge, & Piccolo, 2003) or, though direct evidence is lacking, intrinsic career success (career satisfaction).

3. Although, consistent with the self-serving bias, the causal arrow in Figure 6.1 goes from performance to attributions, much of the research on the relationship between attributions and performance has tended to use attributions or attributional style as a predictor of performance, such that those who tend to make internal attributions perform better than those who tend to make external attributions, in both correlational (Seligman & Schulman, 1986; Silvester, Patterson, & Ferguson, 2003) and experimental (Sharma & Mavi, 2001) studies. For example, Feeley and Foderal (2003) found that midgets' attributions about the causes of their shortness affected their performance in the high jump. On the other hand, evidence does suggest that performance is a source of attributions, especially when individuals are given comparative information (Arnkelsson & Smith, 2000). Although attribution theories would predict that the effect of success on attributions will depend on one's reasoning about the causes of the success (or failure), it stands to reason that the events themselves (in this case, success or performance) will affect attributions. Moreover, in an experimental context, one can design a situation in which either is the cause of the other.

4. It should be noted that individuals also may have tendencies to self-enhance (both positive and negative people engage in processes to improve their self-image); such dual tendencies to self-verify and self-enhance can coexist (Morling & Epstein, 1997).

5. Although state core self-evaluations are shown as the moderating variable in Figure 6.1, it is also possible that trait core self-evaluations be a moderating variable here in the same way as state core self-evaluations. However, for simplicity, we only hypothesize a moderating role of state core self-evaluations, though both can be investigated.

REFERENCES

Abramson, L. Y., Seligman, M. E., & Teasdale, J. D. (1978). Learned helplessness in humans: Critique and reformulation. *Journal of Abnormal Psychology, 87*, 49–74.

Ajzen, I., & Fishbein, M. (1977). Attitude-behavior relations: A theoretical analysis and review of empirical research. *Psychological Bulletin, 84*, 888–918.

Allport, G. W. (1937). *Personality: A psychological interpretation.* New York: Holt, Rinehart, & Winston.

Allport, G. W. (1961). *Pattern and growth in personality.* New York: Holt, Rinehart, & Winston.

Arnkelsson, G. B., & Smith, W. P. (2000). The impact of stable and unstable attributes on ability assessment in social comparison. *Personality and Social Psychology Bulletin, 26*, 936–947.

Bandura, A. (1991). Social cognitive theory of self-regulation. *Organizational Behavior and Human Decision Processes, 50*, 248–287.

Barrick, M. R., & Mount, M. K. (1991). The big five personality dimensions and job performance: A meta-analysis. *Personnel Psychology, 44*, 1–26.

Best, R. G. (2003). *Are self-evaluations at the core of job burnout?* Unpublished doctoral dissertation, Kansas State University.

Bono, J. E., & Judge, T. A. (2003). Core self-evaluations: A review of the trait and its role in job satisfaction and job performance. *European Journal of Personality, 17*, S5–S18.

Bosson, J. K., & Swann, W. B., Jr. (1999). Self-liking, self-competence, and the quest for self-verification. *Personality and Social Psychology Bulletin, 25*, 1230–1241.

Campbell, W. K., & Sedikides, C. (1999). Self-threat magnifies the self-serving bias: A meta-analytic investigation. *Review of General Psychology, 3*, 23–43.

Cattell, R. B. (1965). *The scientific analysis of personality.* Baltimore: Penguin.

Cervone, D., Kopp, D. A., Schaumann, L., & Scott, W. D. (1994). Mood, self-efficacy, and performance standards: Lower moods induce higher standards for performance. *Journal of Personality and Social Psychology, 67*, 499–512.

Cronbach, L. J. (1957). The two disciplines of scientific psychology. *American Psychologist, 12*, 671–684.

Curren, M. T., & Harich, K. R. (1993). Performance attributions: Effects of mood and involvement. *Journal of Educational Psychology, 85*, 605–609.

Di Paula. A., & Campbell, J. D. (2002). Self-esteem in the face of failure. *Journal of Personality and Social Psychology, 83*, 711–724.

Dodgson, P. G., & Wood, J. V. (1998). Self-esteem and the cognitive accessibility of strengths and weaknesses after failure. *Journal of Personality and Social Psychology, 75*, 178–197.

Donovan, J. J., & Williams, K. J. (2003). Missing the mark: Effects of time and causal attributions on goal revision in response to goal-performance discrepancies. *Journal of Applied Psychology, 88*, 379–390.

Earley, P. C., & Lituchy, T. R. (1991). Delineating goal and efficacy effects: A test of three models. *Journal of Applied Psychology, 76*, 81–98.

Eden, D. (1988). Pygmalion, goal setting, and expectancy: Compatible ways to boost productivity. *Academy of Management Review, 13*, 639–652.

Einhorn, H. J., & Hogarth, R. M. (1986). Judging probable cause. *Psychological Bulletin, 99,* 3–19.

Ervin, L. H., & Stryker, S. (2001). Theorizing the relationship between self-esteem and identity. In T. J. Owens, S. Stryker, & N. Goodman (Eds.), *Extending self-esteem theory and research* (pp. 29–55). Cambridge, MA: Cambridge University Press.

Erez, A. (1997). *Core self-evaluations as a source of work-motivation and performance.* Unpublished doctoral dissertation, Cornell University.

Erez, A., & Judge, T. A. (2001). Relationship of core self-evaluations to goal setting, motivation, and performance. *Journal of Applied Psychology, 86,* 1270–1279.

Erez, M., & Zidon, I. (1984). Effect of goal acceptance on the relation of goal difficulty to performance. *Journal of Applied Psychology, 69,* 69–78.

Erikson, E. H. (1959). Identity and the life cycle: Selected papers. *Psychological Issues, 1,* 5–165.

Feick, D. L., & Rhodewalt, F. (1997). The double-edged sword of self-handicapping: Discounting, augmentation, and the protection and enhancement of self-esteem. *Motivation and Emotion, 21,* 147–163.

Finn, J. D., & Rock, D. A. (1997). Academic success among students at risk for school failure. *Journal of Applied Psychology, 82,* 221–234.

Forgas, J. P., Bower, G. H., & Moylan, S. J. (1990). Praise or blame? Affective influences on attributions for achievement. *Journal of Personality and Social Psychology, 59,* 809–819.

Forsyth, D. R., & McMillan, J. (1981). Attributions, affect, and expectations: A test of Weiner's three-dimensional model. *Journal of Educational Psychology, 73,* 393–401.

Ghorpade, J., Hattrup, K., & Lackritz, J. R. (1999). The use of personality measures in cross-cultural research: A test of three personality scales across two countries. *Journal of Applied Psychology, 84,* 670–679.

Heider, F. (1944). Social perception and phenomenal causality. *Psychological Review, 51,* 358–374.

Heider, F. (1958). *The psychology of interpersonal relations.* New York: Wiley.

Heller, D., Judge, T. A., & Watson, D. (2002). The confounding role of personality and trait affectivity in the relationship between job and life satisfaction. *Journal of Organizational Behavior, 23,* 815–835.

Hollenbeck, J. R., & Brief, A. P. (1987). The effects of individual differences and goal origin on goal setting and performance. *Organizational Behavior and Human Decision Processes, 40,* 392–414.

Hollenbeck, J. R., & Klein, H. J. (1987). Goal commitment and the goal-setting process: Problems, prospects, and proposals for future research. *Journal of Applied Psychology, 72,* 212–220.

Hume, D. (1963). *An enquiry concerning human understanding.* New York: Washington Square Press. (Original work published 1748)

Hunter, J. A., Reid, J. M., Stokell, N. M., & Platow, M. J. (2000). Social attribution, self-esteem, and social identity. *Current Research in Social Psychology, 5,* 97–125.

James, W. (1890). *The principles of psychology.* Available: http://psychclassics.yorku.ca/James/Principles/.

Jones, E. E., & Davis, K. E. (1965). From acts to dispositions: The attribution process in person perception. In L. Berkowitz (Ed.), *Advances in experimental social psychology* (Vol. 2). Orlando, FL: Academic Press.

Judge, T. A., & Bono, J. E. (2001). Relationship of core self-evaluations traits—self-esteem, generalized self-efficacy, locus of control, and emotional stability—with job satisfaction and job performance: A meta-analysis. *Journal of Applied Psychology, 86,* 80–92.

Judge, T. A., & Larsen, R. J. (2001). Dispositional source of job satisfaction: A review and theoretical extension. *Organizational Behavior and Human Decision Processes, 86,* 67–98.

Judge, T. A., Erez, A., & Bono, J. E. (1998). The power of being positive: The relationship between positive self-concept and job performance. *Human Performance, 11,* 167–187.

Judge, T. A., Erez, A., Bono, J. E., & Thoresen, C. J. (2002). Are measures of self-esteem, neuroticism, locus of control, and generalized self-efficacy indicators of a common core construct? *Journal of Personality and Social Psychology, 83,* 693–710.

Judge, T. A., Locke, E. A., & Durham, C. C. (1997). The dispositional causes of job satisfaction: A core evaluations approach. *Research in Organizational Behavior, 19,* 151–188.

Judge, T. A., Locke, E. A., Durham, C. C., & Kluger, A. N. (1998). Dispositional effects on job and life satisfaction: The role of core evaluations. *Journal of Applied Psychology, 83,* 17–34.

Kammeyer-Mueller, J. D., Judge, T. A., & Piccolo, R. F. (2003). Self-esteem and occupational success: Test of a dynamic model. Unpublished manuscript.

Kant, I. (1998). *The critique of pure reason* (P. Guyer & A. W. Wood, Eds. & Trans.). New York: Cambridge University Press. (Original work published 1781)

Kelley, H. H. (1967). Attribution in social psychology. *Nebraska Symposium on Motivation, 15,* 192–238.

Kelley, H. H. (1973). The processes of causal attribution. *American Psychologist, 28,* 107–128.

Leibniz, G.W. (1982). *New essays on human understanding.* (P. Remnant & J. Bennett, Eds. & Trans.). New York: Cambridge University Press. (Original work published 1765)

Locke, E. A. (1968). Toward a theory of task motivation and incentives. *Organizational Behavior and Human Performance, 3,* 157–189.

Locke, E. A. 1997. The motivation to work: What we know. *Advances in Motivation and Achievement, 10,* 375–412.

Locke, E. A., & Latham, G. P. 1990. *A theory of goal setting and task performance.* Englewood Cliffs, NJ: Prentice-Hall.

Locke, E. A., Shaw, K. N., Saari, L. M., and Latham, G. P. (1981). Goal setting and task performance: 1968–1980. *Psychological Bulletin, 90,* 125–152.

Loehlin, J. C. (1992). *Genes and environment in personality development.* Newbury Park, CA: Sage.

Loehlin, J. C., McCrae, R. R., Costa, P. T., & John, O. P. (1998). Heritabilities of common and measure-specific components of the Big Five personality factors. *Journal of Research in Personality, 32,* 431–453.

Martinko, M. J. (1995). The nature and function of attribution theory within the organizational sciences. In M. J. Martinko (Ed.), *Attribution theory: An organizational perspective* (pp. 7–16). Delray Beach, FL: St. Lucie Press.

Martinko, M. J., & Thomson, N. F. (1998). A synthesis and extension of the Weiner and Kelley attribution models. *Basic and Applied Social Psychology, 20,* 271–284.

Mayer, J. D., Gaschke, V., Braverman, D. L., & Evans, T. W. (1992). Mood-congruent judgment is a general effect. *Journal of Personality and Social Psychology, 63,* 119–132.

McAdams, D. P. (1990). Unity and purpose in human lives: The emergence of identity as a life story. In A. I. Rabin, R. A. Zucker, R. A. Emmons, & S. Frank, *Studying persons and lives* (pp. 148–200). New York: Springer-Verlag.

Mento, A. J., Steel, R. P., & Karren, R. J. 1987. A meta-analytic study of the effects of goal setting on task performance: 1966–1984. *Organizational Behavior and Human Decision Processes, 39,* 52–83.

Morelli, G., Krotinger, H., & Moore, S. (1979). Neuroticism and Levenson's locus of control scale. *Psychological Reports, 44,* 153–154.

Morling, B., & Epstein, S. (1997). Compromises produced by the dialectic between self-verification and self-enhancement. *Journal of Personality and Social Psychology, 73,* 1268–1283.

Murray, H. A. (1938). *Explorations in personality.* New York: Oxford University Press.

Needles, D. J., & Abramson, L. Y. (1990). Positive life events, attributional style, and hopefulness: Testing a model of recovery from depression. *Journal of Abnormal Psychology, 99,* 156–165.

Oxford University Press. (2003). *Oxford English Dictionary OED Online.* Oxford: Oxford University Press.

Phillips, J. M., & Gully, S. M. (1997). Role of goal orientation, ability, need for achievement, and locus of control in the self-efficacy and goal-setting process. *Journal of Applied Psychology, 82,* 792–802.

Revelle, W. (1995). Personality processes. *Annual Review of Psychology, 46,* 295–328.

Roberts, B. W., & DelVecchio, W. F. (2000). The rank-order consistency of personality traits from childhood to old age: A quantitative review of longitudinal studies. *Psychological Bulletin, 126,* 3–25.

Rode, J. C. (2002). *The role of core evaluations within a comprehensive job and life satisfaction model: A longitudinal analysis.* Unpublished doctoral dissertation, Indiana University.

Rotter, J. B. (1966). Generalized expectancies for internal versus external control of reinforcement. *Psychological Monographs: General and Applied, 80,* 1–28.

Sedikides, C. (1992). Changes in the valence of the self as a function of mood. In M.S. Clark *Emotion and social behavior: Review of personality and social psychology* (Vol. 14, pp. 271–311). Newbury Park, CA: Sage.

Sedikides, C. (1993). Assessment, enhancement, and verification determinants of the self-evaluation process. *Journal of Personality and Social Psychology, 65,* 317–338.

Sedikides, C. (1995). Central and peripheral self-conceptions are differentially influenced by mood: Tests of the differential sensitivity hypothesis. *Journal of Personality and Social Psychology, 69,* 759–777.

Seligman, M. E. P., & Schulman, P. (1986). Explanatory style as a predictor of productivity and quitting among life insurance agents. *Journal of Personality and Social Psychology, 50,* 832–838.

Sharma, V., & Mavi, J. (2001). Self-esteem and performance on word tasks. *Journal of Social Psychology, 141,* 723–729.

Silvester, J., Patterson, F., & Ferguson, E. (2003). Comparing two attributional models of job performance in retail sales: A field study. *Journal of Occupational and Organizational Psychology, 76,* 115–132.

Snyder, C. R. (2002). Hope theory: Rainbows in the mind. *Psychological Inquiry, 13,* 249–275.

Spielberger, C. D., Gorsuch, R. L., Lushene, R., Vagg, P. R., & Jacobs, G. A. (1983). *Manual for the State-Trait Anxiety Inventory (Form Y).* Palo Alto, CA: Consulting Psychologists Press.

Steyer, R., Ferring, D., & Schmitt, M. J. (1992). States and traits in psychological assessment. *European Journal of Psychological Assessment, 8,* 79–98.

Swann, W. B., Stein-Seroussi, A., & Giesler, R. B. (1992). Why people self-verify. *Journal of Personality and Social Psychology, 62,* 392–401.

Thomas, K. M., & Mathieu, J. E. (1994). Role of causal attributions in dynamic self-regulation and goal processes. *Journal of Applied Psychology, 79,* 812–818.

Tubbs, M. E. (1986). Goal setting: A meta-analytic examination of the empirical evidence. *Journal of Applied Psychology, 71,* 474–483.

Wambach, R. L., & Panackal, A. A. (1979). Age, sex, neuroticism, and locus of control. *Psychological Reports, 44,* 1055–1058.

Watson, D., Clark, L.A., & Tellegen, A. (1988). Development and validation of brief measures of positive and negative affect: The PANAS scales. *Journal of Personality and Social Psychology, 54,* 1063–1070.

Weiner, B. (1980). *Human motivation.* New York: Holt, Rinehart & Winston.

Weiner, B. (1987). The social psychology of emotion: Applications of a naive psychology. *Journal of Social and Clinical Psychology, 5,* 405–419.

Weiner, B. (1990). Attribution in personality psychology. In L. A. Pervin (Ed.), *Handbook of personality: Theory and research* (pp. 465–485). New York: Guilford Press.

Weiner, B., & Graham, S. (1999). Attribution in personality psychology. In L. A. Pervin & O. P. John (Eds.), *Handbook of personality: Theory and research* (2nd ed., pp. 605–628). New York: Guilford Press.

Weiner, B., & Russell, D., & Lerman, D. (1978). Affective consequences of causal ascriptions. In J. H. Harvey, W. J. Ickes, & R. F. Kidd (Eds.), *New directions in attribution research* (Vol. 2, pp. 59–91). Hillsdale, NJ: Erlbaum.

Winefield, A. H., Tiggemann, M., & Winefield, H. R. (1992). Unemployment distress, reasons for job loss and causal attributions for unemployment in young people. *Journal of Occupational and Organizational Psychology, 65,* 213–218.

Wright, P. M. (1990). Operationalization of goal difficulty as a moderator of the goal difficulty–performance relationship. *Journal of Applied Psychology, 75,* 227–234.

CHAPTER 7

AN EXPLORATORY STUDY
OF WORKPLACE AGGRESSION

Mark J. Martinko
Florida State University

Sherry E. Moss
Florida International University

ABSTRACT

This chapter reports on differences in attributions for workplace events for a
sample of inmates incarcerated for violent offenses versus a sample of nonin-
carcerated MBA students. The results indicate that, as predicted, the attribution styles of the violent inmates were significantly more narcissistic than the
attribution styles of the MBA sample. The role of attributions and attributional styles in predicting incidents of workplace aggression is discussed.

In recent years, interest in the area of organizational aggression has been
heightened by popular media accounts that describe sensational acts of
violence in organizations (e.g., Elliott & Jarrett, 1994; Filipczak, 1993; Silverstein, 1994; Stuart, 1992; Toufexis, 1994). These accounts frequently
mention organizational situations such as poor working conditions and
oppressive supervision, which appear to precipitate acts of aggression
(e.g., Armour, 1998; Elliott & Jarrett, 1994; Grimsley, 1998; Stuart, 1992;
Toufexis, 1994; Yates, 1995), but tend to focus on describing the traits and

Attribution Theory in the Organizational Sciences, pages 133–150

characteristics of the "typical" perpetrator (Folger & Skarlicki, 1998). Thus, for example, Toufexis (1994) describes perpetrators of organizational aggression as being focused on their jobs, having few outside interests, and having extremist opinions. Similarly, Stuart (1992) describes likely aggressors as having high job identification and few relationships outside of their jobs.

In response to the needs illuminated by the practitioner literature, there has also been a significant increase in both theory and research concerned with dysfunctional organizational behavior. O'Leary-Kelly, Griffin, and Glew (1996) provided one of the first theoretical descriptions of this area and concentrated primarily on identifying organizational factors related to the frequencies and probabilities of aggressive incidents. Much of the theory and research that followed has concentrated on organizational factors associated with dysfunctional organizational behavior. More specifically, the theoretical discussions of organizational aggression presented by Neuman and Baron (1998), Baron and Richardson (1994), and Folger and Skarlicki (1998) have also focused on organizational factors that are likely to cue aggression. Moreover, several of these perspectives also appear to suggest that perspectives that emphasize individual differences are unlikely to be successful in explaining the causes of organizational aggression. Thus, for example, O'Leary-Kelley and colleagues (1996) indicate that "predictions of individual violence tend to be greatly overestimated" and appear to discourage the investigation of individual difference factors. Similarly, Folger and Skarlicki suggest that earlier accounts of incidents of organizational violence tended to overestimate the role of individual differences and underestimate the power of the situation. Finally, in a discussion of deviant organizational behavior (which includes aggression), Robinson and Greenberg (1998) asserted that "no clear picture emerges of the 'deviant personality type' in organizations" and suggested that personality variables account for only a minor portion of the variance in deviant workplace behavior. Thus, the conceptual literature appears to suggest that the potential of individual difference variables in explaining workplace aggression is limited, particularly if they are considered as independent predictors of workplace aggression.

In general, the empirical research on organizational aggression tends to emphasize the role of organizational factors in stimulating organizational aggression and dysfunctional behaviors (e.g., Greenberg, 1990; Robinson & O'Leary-Kelley, 1998; Skarlicki & Folger, 1997). Although these studies have found that organizational-level factors do make significant contributions to the variance in aggressive and deviant workplace behaviors, the majority of the variance in most of these studies is unexplained, suggesting that factors other than the ones included in these studies also play a significant role in the emergence of dysfunctional behaviors such as organizational aggression.

Recently, theories and models of organizational aggression have emerged that recognize the role of individual differences (Aquino, Grover, Bradfield, & Allen, 1999; Martinko & Zellars, 1998; Neuman & Baron, 1998). The most detailed and specific of these models is the cognitive appraisal perspective presented by Martinko and Zellars (1998), which asserts that although organizational-level variables are important, individual differences in member's cognitive appraisals (i.e., attributions) regarding personal work-related outcomes play a central role in predicting whether or not an individual is motivated to aggress. The model proposes that understanding the ways in which aggressive versus nonaggressive individuals appraise information about organizational events is an important factor in understanding and predicting organizational aggression. More specifically, the model proposes that violent versus nonviolent individuals can be differentiated by attribution styles such that individuals oriented toward organizational aggression and violence are likely to view negative organizational events as caused by external and stable factors such as a punitive supervisor whereas less violent individuals are more likely to attribute negative organizational events to internal or external unstable factors such as the economy or their own shortcomings. Another recent paper by Neuman and Baron (1998) suggests a similar approach, noting that differences in individuals' cognitive appraisals of workplace incidents may account for at least some of the differences in individual reactions to negative outcomes in the workplace.

There is already some support in the organizational literature for the notion that individual differences in the cognitive appraisal process may help explain why some individuals react aggressively whereas others are nonaggressive. Most recently, Douglas and Martinko (2001) found that attribution styles accounted for a significant proportion of the variability in reports of workplace aggression. In addition, earlier research by Storms and Spector (1987) found that individuals with an external locus of control were more likely to engage in acts of industrial sabotage than individuals with an internal locus of control. Similar results were found by Allen and Greenberger (1980) who found that individuals who perceived that they had relatively little control over important work-related outcomes were more likely than others to engage in acts of organizational vandalism. Finally, Spector (1982) found that males with an external locus of control were more likely than others to use aggressive behaviors in managing others.

It should be noted that although there has been very little research relating attributions to organizational aggression, the broader research on aggression supports the notion that hostile attribution biases are related to aggression. Indeed, work on the hostile attribution bias (Dodge, 1987; Dodge, Murphy, & Buchsbaum, 1984; VanOostrum, 1997) indicates that attributions of harmful intent are strongly related to aggressive retaliation. More specifically, research by Nasby, Hayden, and DePaulo (1967) and Dodge and Coie (1987) found that subjects who attribute hostile intent to the actions of others are likely to be aggressive, even when the actions of the others are ambiguous. Other studies verify that when negative outcomes are attributed to actions that are perceived to be external, controllable, and/or intentional, anger often results (Averill, 1982; Betancourt & Blair, 1992; Weiner, 198, 19952; Weiner, Graham, & Chandler, 1982). Thus, based on theory and research from the organizational and social psychological literature, there is strong support for the notion that attribution styles will be related to the incidence of workplace aggression.

In addition, work on sociopathology and narcissism has suggested that perpetrators of socially deplorable acts are often self-centered and narcissistic (Kets de Vries, 1993; Maccoby, 2003; Zimbardo, 1985). More specifically, sociopaths and narcissists see themselves as the center of their world and are unable to empathize with and understand the behavior of others. Because they see themselves as the center of their worlds, we believe they are best characterized by an accentuated self-serving bias, taking credit for their successes (i.e., making internal attributions) and blaming their failures on external factors (i.e., making external attributions). While the notion of blaming external factors is consistent with the hostile attribution bias, the work on aggression appears to be silent regarding the ways in which hostile individuals process information about success. We believe that the work on sociopathology and narcissism helps fill this void, suggesting that such individuals would most likely make internal, and probably stable, attributions for their successes. Thus we believe that perpetrators of organizational violence probably display accentuated self-serving biases, taking credit for success while blaming failure on external factors.

Despite both theoretical and empirical support for the notion that attributional processes are related to incidents of organizational aggression, with the exception of the few studies mentioned above, there has been no direct validation of an attributional perspective of organizational aggression. At least one of the reasons why such a validation study has not been reported is logistical problems. An ideal study would compare the attributions of individuals who have perpetrated acts of organizational aggression with a nonaggressive control sample with similar demographic and occupational histories. Another reason why such a validation study had not been reported is that the actual occurrence of violent incidents is infrequent (Elliott & Jarrett, 1994). Moreover, gaining access to individuals who have perpetrated such acts would not be easy. If access could be obtained, there would be a potential problem with respondent bias since individuals accused of crimes would probably not want to provide candid data that could result in incrimination. Finally, at least at the current time, criminal justice data systems do not record acts of organizational aggression in a separate category. Thus on a statewide or nationwide basis, there is no efficient system for identifying individuals who have perpetrated acts of organizational aggression.

Although the logistical problems described above may be able to be overcome with sufficient resources, it would seem that a general exploratory test of the model would be appropriate until a more rigorous test can be done. More specifically, if attribution styles are related to acts of organizational aggression, and if organizational aggression is a subset of general aggression, one would expect that individuals who are predisposed to more general forms of aggression would have attributional styles that are similar to those believed to be related to organizational aggression. In particular, as suggested above, we would expect that:

Hypothesis 1:   *There will be a difference in the attributional styles for work-related events of a known sample of aggressive individuals as compared to a sample of normal individuals such that the aggressive sample will make more "narcissistic" (internal and stable) attributions for success than the normal sample.*

Hypothesis 2:   *There will be a difference in the attributional styles for work-related events of a known sample of aggressive individuals as compared to a sample of normal individuals such that the aggressive sample will make more "hostile" attributions (external and stable) for failures than the normal sample.*

We do not know of any similar comparison that has been done between samples of known violent individuals versus a control population.

METHODS

Participants

Twenty-one male inmates residing in a state prison and 78 students in four different MBA programs voluntarily responded to the Organizational Attribution Style Questionnaire II.

Inmate Sample

The prison sample consisted of 14 white, one Hispanic, and six black inmates ranging in age from 31 to 61 with an average of 38.56 years. Almost all were serving time for violent crimes including first-degree murder, second-degree murder, third-degree murder, kidnapping, and robbery. Although only inmates convicted of violent crimes were requested for the sample, at least two criminal histories suggested that the offenses may not have been violent (e.g., robbery without a firearm or deadly weapon). However, the remainder of the sample had clearly been convicted of violent crimes and it is likely that violence could have also been associated with the crimes of the two other inmates, although more detailed criminal histories were not available.

MBA Samples

The first MBA sample consisted of 14 (seven white, six Hispanic, and one Asian) students ranging in age from 22 to 33 with an average age of 25.43. Nine were male and five female. The second sample consisted of 17 students from an executive MBA program ranging in age from 29 to 44 with an average age of 35.18. Three were females and 14 were males. Seven were white, three were black, and six were Hispanic. The third sample consisted of 18 executive MBA students ranging in age from 27 to 54 with an average age of 36.83. Seven participants were females and 11 were males. Three were white, two were black, 12 Hispanic, and one Asian. The fourth MBA sample included 28 black students enrolled in a joint undergraduate/MBA program. Their ages ranged from 19 to 24 with an average age of 21.21. Thirteen were females and 15 were males. All of the executive MBA students were currently employed and more than 90% of the other MBA students were employed or had had significant work experience. Table 7.1 summarizes the demographics of the five samples.

Table 7.1.   Sample Demographics

	Prison	MBA1*	MBA2*	MBA3*	MBA4**
Age					
Average Age	38.56	25.43	35.18	36.83	21.21
High Age	61	33	44	54	24
Low Age	31	22	29	27	19
Gender					
# Males	21	9	14	11	15
# Females	0	5	3	7	13
Ethnicity					
# Whites	14	7	7	3	0
# Blacks	6	0	3	2	28

Table 7.1.   Sample Demographics (Cont.)

	Prison	MBA1*	MBA2*	MBA3*	MBA4**
# Hispanics	1	6	6	12	0
# Asians	0	1	0	1	0

*MBA samples 1, 2, and 3 were collected at a university in a region with a large Hispanic population.
**MBA sample 4 was collected at a historically black university.

Instrumentation

The questionnaire administered was the Organizational Attribution Style Questionnaire, Form II (OASQ II), which is a revised version of the Organizational Attribution Style Questionnaire (Campbell & Martinko, 1998; Kent & Martinko, 1995a; Martinko, 1994). The OASQ consists of eight workplace scenarios (four positive and four negative). An example of a positive scenario is, "You receive an unusually high raise compared to others in your department." An example of a negative scenario is, "You receive a poor performance evaluation." Respondents are asked to read each scenario and imagine themselves in the situation. They are then asked four questions about attributional explanations and two questions measuring attributional dimensions. First, they are asked to describe the extent to which the outcome of the scenario was caused by (a) their ability, (b) their efforts, (c) the situation and circumstances, and (d) luck/chance. Their responses to these attributional explanations are measured on a seven-point Likert scale ranging from "Did not determine outcome at all" to "Totally determined outcome."

A second measure of attribution style was composed of items measuring attributional dimensions. Following each scenario, the subjects were asked to respond to one item measuring *locus of causality* and one item measuring *stability*. The locus item asked, (1) "To what extent is the cause (or causes) of the outcome due to something about you or due to other people or circumstances?" and subjects respond on a seven-point Likert scale ranging from "Completely due to me" to "Completely due to other people or circumstances." The stability item asked, (2) "To what extent is the cause (or causes) of the outcome described above?" "stable over time" to "variable over time," as indicated on a seven-point Likert scale.

The OASQ II differs from earlier versions of the OASQ in that it provides separate scales to measure attributional explanations as well as attributional dimensions. Earlier research and discussions (Kent & Martinko, 1995a, 1995b) regarding the OASQ noted that the philosophy that was used in constructing most attribution scales was to measure attributional dimensions since it was assumed that it was inappropriate to infer dimensions from attributional explanations (Russell, McAuley, & Tarico, 1987). However, research has demonstrated that attributional explanations can be reliably recorded (Liden & Mitchell, 1985). In addition, qualitative interviews with informants suggest that they often use attributional explanations rather than referring to attributional dimensions when they explain key organizational outcomes (Campbell & Martinko, 1998). As a result, the OASQ II reflects both explanations and dimensions and is intended as an exploratory questionnaire to help determine whether measuring attributional explanations or dimensions provides the most robust results. Thus, a research question that was of secondary interest was whether measures of attributional explanations or measures of attributional dimensions did a better job of distinguishing between the two groups.

Finally, it should be noted that the OASQ II measures only two attributional dimensions: locus of causality and stability. Prior research has indicated that locus of causality, controllability, and intentionality are highly intercorrelated, suggesting that separate measures relating to each of these are redundant (Kent & Martinko, 1995a, 1995b). In addition, the stability measure is often confounded with the globality dimension (Kent & Martinko, 1995a), so only a measure of stability is used. As Weiner (1985, 1995) has noted, attributional dimensions are highly dependent on the domain of study such that one does not necessarily expect cross-situational consistency in attributional dimensions when the behavioral domains are very different. In recognition of the potential problems in trying to measure too many dimensions within the same questionnaire, the OASQ II limits both attributional explanations and dimensions to those that have been found to appear most frequently and reliably across behavioral domains.

The dependent variables generated from the OASQ were locus of causality and stability for positive and negative events (using dimensions) and measures of attributions to ability, effort, situation/circumstances, and luck/chance for positive and negative events (using explanations).

Procedure

In the case of the inmates, all of the individuals in the study were identified beforehand through prison records and asked to voluntarily participate in the study. Seven individuals who were identified beforehand did not participate because they were released or in detention. Seven other individuals were recruited by one of the prison administrators and voluntarily agreed to participate in the study.

One of the authors administered the questionnaire in the prison library during the time period immediately after the inmates' lunch. A prison administrator and the prison librarian were also present. The researcher distributed and then collected voluntary consent forms. All inmates completed and signed the forms. The researcher then distributed the OASQ II asking the inmates to relate the incidents on the form to their current job in the prison (all were assigned some job) or a job they once held. They were asked to make reasonable assumptions regarding incidents such as pay raises and promotions, for example, "imagining that you could receive a raise in your current job." All of the inmates completed the OASQ II within 45 minutes.

The OASQ II was distributed at the beginning of each of the MBA classes and the participants were asked to voluntarily complete the form using their current job or prior jobs as a point of reference. The return rate for the MBA classes was approximately 60 per cent.

RESULTS

Scale Reliabilities

A scale was created for each attributional explanation (e.g., ability) across the four positive events and then across the four negative events. Thus, there were two scales for ability (positive events, negative events), two for effort, two for situation and circumstances, and two for luck/chance. A scale was also created for each attributional dimension (locus of causality, stability) across the four positive and the four negative events. The alpha coefficients for all of the scales are reported in Table 7.2. All reliability coefficients for both attributional explanations and attributional dimensions were in an acceptable range (i.e., greater than .70; Nunnally, 1970), with the exception of the attributional explanation of situation/circumstances, which had an alpha of .69 for positive events and an alpha of .57 for negative events.

Preliminary Analyses

Before proceeding with the main analysis, the demographic composition of the five subsamples was analyzed to determine which variables needed to be statistically controlled in subsequent analyses. Since all MBA

Table 7.2.   Coefficient Alphas for All Scales

	Alpha	*N of items*
Attributional Explanations		
Positive Events		
Ability	.76	4
Effort	.77	4
Situation/Circumstances	.69	4
Luck/Chance	.90	4
Negative Events		
Ability	.76	4
Effort	.79	4
Situation/Circumstances	.57	4
Luck/Chance	.80	4
Attributional Dimensions		
Positive Events		
Locus of Control	.70	4
Stability	.85	4
Negative Events		
Locus of Control	.71	4
Stability	.75	4

samples were to be combined and compared to the prison sample, initial analyses compared MBAs versus inmates. The first ANOVA revealed that the prison population was significantly older than the MBA samples combined (F 1, 93 = 21.521, $p < .001$). The next two analyses were chi-square analyses examining differences in gender and ethnicity for the prison and MBA samples. Clearly, when the MBA samples are grouped and compared to the prison sample, there are significant differences in both ethnicity (Pearson chi-square = 10.69, $p < .001$) and gender (Pearson chi-square = 16.14, $p < .001$). Thus, for all remaining analyses, age, gender, and ethnicity were used as covariates.

Main Analyses

Multivariate analysis of covariance (MANCOVA) was the main analytical tool. Using the combined MBA sample and the inmate sample as the comparison groups, the differences in both attributional explanations (ability, effort, situation/circumstances, luck/chance) and attributional dimensions (locus of causality and stability) for both positive and negative events were examined while statistically controlling for age, ethnicity, and gender. This method of analysis allows us to test both H1 and H2 using attributional dimensions and explanations as separate measures of subject attributions.

The first MANCOVA examined differences between violent prisoners and MBAs on attributional *dimensions* (locus of causality, stability) for *positive events*. The multivariate test statistic, Wilk's lambda, was significant (F 2, 88 = 5.151, $p < .008$). Follow-up univariate analyses revealed significant effects for both locus of causality (F 1, 89 = 8.106, $p < .005$) and stability (F 1, 89 = 7.695, $p < .007$). The means reported in Table 7.3 indicate that the prisoners were both more internal and stable in their attributions for positive events. This finding supports H1, which states that inmates would be more likely to have a narcissistic tendency to take credit for positive outcomes.

Table 7.3.   Adjusted Means for Locus of Causality and Stability

		Positive Events Mean[a]	Negative Events Mean
Locus			
	Prison	5.97	3.15
	MBA	5.10	3.77
Stability			
	Prison	5.57	3.30
	MBA	4.34	3.44

[a] Higher numbers indicate more internal and stable attributions.

The second MANCOVA examined differences between violent prisoners and MBAs on attributional *explanations* (ability, effort, etc.) for *positive events*. The multivariate test statistic, Wilk's lambda, was significant (F 4, 86 = 3.789, $p < .007$). Univariate F-tests indicated a significant difference between the two groups on luck/chance (F 1, 93 = 11.74, $p < .001$) and marginally significant differences on ability (F1, 93 = 3.29, $p < .073$) and effort (F 1, 93 = 3.68, $p < .058$). Examination of Table 7.4 indicates that the prisoners were more likely to attribute their positive outcomes to ability and effort and less to luck/chance. Though marginal, these findings also support the idea that prisoners possess an extreme (narcissistic) version of the self-serving bias for positive outcomes. Furthermore, they clearly discount the possibility that luck or chance could have played a role in their successes.

Table 7.4.   Adjusted Means for Attributional Explanations

		Positive Events	Negative Events
		Mean	Mean
Ability	Prison	6.26	2.32
	MBA	5.75	3.60
Effort	Prison	6.32	2.62
	MBA	5.80	3.58
Situation	Prison	4.28	4.93
	MBA	4.47	4.95
Luck/Chance	Prison	1.54	1.70
	MBA	2.89	2.92

The third MANCOVA examined differences between violent prisoners and MBAs on attributional *dimensions* for *negative events*. The multivariate test statistic, Wilk's lambda, was not significant. Thus, H2 was not supported for attributional dimensions.

The fourth MANCOVA examined differences between groups on attributional *explanations* for *negative events*. The multivariate test statistic, Wilk's lambda, was significant (F 4, 86 = 6.33, $p < .0001$). Examination of the univariate results reveals significant differences between the inmates and the MBAs on ability (F 1, 89 = 10.43, $p < .002$), effort (F 1, 89 = 4.85, $p < .03$), and luck (F 1, 89 = 9.87, $p < .002$). Examination of the means in Table 7.4 indicates that the inmates were less likely to attribute negative outcomes to their own ability or effort, and were also less likely to attribute them to luck/chance. This finding is only partially supportive of H2. Attributions to luck/chance are external but unstable.

DISCUSSION

The results confirmed the notion that there would be significant differences in the ways in which violent versus nonviolent individuals make attributions regarding causality for workplace events. Specifically, both measures of attributions (dimensions and explanations) provided support for the hypothesis that prisoners would display an exaggerated form of the self-serving bias for positive outcomes. They clearly took credit for their successes and discounted external explanations.

The results for negative events were not as unequivocal. Although the pattern of results for *dimensions* found no significant differences between aggressive prisoners and MBAs for negative events, the pattern of results for *explanations* revealed that the prisoners attributed failure significantly less to ability and effort than the control group. Thus, although the prison sample did not demonstrate the prototypical hostile attribution bias (external and stable explanations), their biases can be described as self-serving in they were significantly less likely than the control group to attribute their failures to ability or effort (i.e., internal explanations).

There are several possible explanations for the inability to confirm H2, the notion that hostile attribution biases regarding failure would be stronger in the prison sample. One potential explanation concerns the wording of our questionnaire. The alternative on the questionnaire representing external and stable attributions was worded "situation/circumstances" rather than a more specific alternative such as "another person." This alternative should probably be added in future studies. In addition, the finding that items measuring this attribution also had slightly lower than acceptable scale reliabilities is another reason we were not able to identify significant differences and suggests that this scale needs to be revised.

It is also possible that the relationships between attribution style and the tendency to engage in violent behavior may not be as direct as our theory suggests. More specifically, it is possible that individuals with narcissistic styles, characterized by accentuated self-serving biases, taking credit for success, and blaming external causes for failure, also have weak self-esteem. As a result, they may be more likely than others to interpret outcomes as failures and, when there are clear external causes to which they can attribute their failures, they may be more likely than others to blame external and stable causes. Such an interpretation would be consistent with Aquino and colleagues' (1999) work that demonstrates that perceived victimization often leads to aggression. It is also consistent with Weiner's (1995) and Martinko and Thomson's (1998) assertions that the realities of situations are the primary determinant of attributions and that attribution styles are only likely to have an impact when situations are ambiguous. Thus, it is possible that although the tendency of individuals with narcissism is to attribute failure to external and unstable situations, they may also be more likely to react to perceived threats and make external and stable attributions when situational cues are salient.

Finally, as indicated earlier, we are well aware that this study does not provide an optimal test of whether or not there are differences in the attributional styles of employees who are prone to aggression versus those who are less aggressive. The OASQ II provides hypothetical situations and it can be argued that the inmates may not have been able to relate to the hypothetical situations as well as the control sample. In addition, the inmates were incarcerated and most had already served more than a year of prison time, so it is possible that their aggressive tendencies had been affected by rehabilitation. Thus, we simply do not know how an individual who is currently disposed to aggression and violence would respond. In addition, although the majority of the inmates in our sample were incarcerated for violent crimes, these were not necessarily perpetrated in organizational contexts. Thus, our sample is less than optimal and there may be important differences between this sample and the population of individuals who have actually committed acts of organizational violence and aggression. It can also be argued that a better demographic match could have been made between the inmate sample and the control group. Finally, because there were some differences in the rate at which the two groups voluntarily participated in the study, there is a possibility that the differences between the two groups are a result of the selection biases. Although this possibility does not seem likely, additional confirming data would eliminate doubts about the results and data set.

Despite the limitations that we acknowledged above, the results clearly demonstrate that the attribution styles regarding workplace events of individuals who are known to have perpetrated violent acts are significantly different than the attribution styles of a more typical sample. These findings are mostly consistent with those of Douglas and Martinko (2001) as well as the results of the studies of Storms and Spector (1987), Allen and Greenberger (1980), and Spector (1982), which indicate that specific types of attributions are related to aggressive behaviors. Thus, the results are encouraging and suggest that further research is warranted. Clearly, it would be desirable to replicate this study with a sample of inmates incarcerated for acts of *organizational* aggression and violence. It would also be desirable to conduct qualitative interviews to more fully understand how the target sample differs in processing information about causality versus a control sample. Although we proposed such a study as part of the current study, we found that there were a number of practical and logistical obstacles that could not be overcome in order to gain the type of access needed to conduct such in-depth interviews in the prison sample. Future investigators need to be aware of and be prepared to overcome the limitations of conducting research in prison systems. External funding to pay for the additional staffing needed to conduct such studies is one possible means of gaining additional access to data.

The results also raise some theoretical issues that warrant further research. The finding that the majority of the differences between the inmate sample and the control sample could be accounted for by reactions to the positive rather than the negative events warrants some consideration. There is no apparent explanation for this finding. One potential explanation is that it was simply an artifact of the samples. Another interesting possibility is that the results are biased because of impression management (Gardner & Martinko, 1988). More specifically, conversations with various law enforcement and prison officials reminded us that the term "con" suggests that incarcerated individuals are less truthful and often intentionally more deceiving than people in the normal population. Thus, as one informant from law enforcement suggested, "tongue-in-cheek," interviewing the inmates would reveal that virtually all of them were wrongfully incarcerated. Although we are not necessarily suggesting that this perspective is valid, there is the possibility that the inmates were less candid about some of their data than the control sample and this could explain why they did not appear to react as strongly to the negative events.

Another interesting finding from a theoretical perspective is the relatively large difference between the attributions for negative versus positive events in the prison sample versus the control sample (see Tables 7.3 and 7.4). This finding clearly indicates that, at least in this study, the self-serving bias was considerably more pronounced in the inmate versus the control sample. The notion of an accentuated self-serving bias is consistent with depictions of sociopathic personalities that tend to prey on others and fail to accept responsibility for their own behavior (Kets de Vries, 1993; Zimbardo, 1985). Again, this finding would be particularly interesting to pursue in future research.

Finally, in some respects the results from our study appear to contradict those of Allen and Greenberger (1980) and Storms and Spector (1987) who found that an external locus of control was related to acts of vandalism and sabotage, respectively. Potential explanations for the differences in findings are differences in the constructs of locus of control versus attribution style and differences in the dependent variables in the respective studies. More specifically, locus of control is a measure of general life orientation whereas attribution style and the construct of locus of causality relate to perceptions about success and failure in achievement situations. The measures of locus of control do not separate reactions to success and failure incidents whereas the OASQ identifies to both positive and negative events. Thus, there are dissimilarities in both the constructs and measures of attribution style and locus of control. In addition, when placed on a continuum of aggressiveness, the behaviors of vandalism and sabotage are less extreme forms of aggression than the violent acts for which the prisoners had been incarcerated. Thus, it may be that the attribution styles of individuals who perpetrate acts of aggression intended to do bodily harm are different than those who choose less personalized forms of aggression such as sabotage and vandalism. In the future, researchers should be attentive to differences in the forms of aggression and be aware that different individual traits may be associated with different forms of aggression.

CONCLUSIONS

The results clearly demonstrated that there were significant differences in the attributions for workplace events between a sample of violent inmates and a control sample. This study represents an important contribution to understanding aggressive behavior in organizations in that it compares a known sample of violent individuals with a more typical sample as opposed to attempting to measure aggression by self-reports (e.g., Baron & Neuman, 1996; Robinson & O'Leary-Kelley, 1998). Although there were limitations regarding the sample and methodology of the current study, the results appear robust enough to warrant further investigation of the relationships between attribution style and incidents of workplace aggression.

REFERENCES

Aquino, K., Grover, S. L., Bradfield, M., & Allen, D. G. (1999). The effects of negative affectivity, hierarchical status, and self-determination on workplace victimization. *Academy of Management Journal, 42*(3), 260–272.

Allen, V. L., & Greenberger, D. B. (1980). Destruction and perceived control. In A. Baum & J. E.

Singer (Eds.), *Applications of personal control.* Hillsdale, NJ: Erlbaum.

Armour, S. (1998, September 9). Running of the bullies trampling workplace morale andproductivity: Offenders can spread ill will from top down. *USA Today,* Money Section, p. 1B.

Averill, J. R. (1979). Anger. In R. A. Dienstbier (Ed.), *Nebraska Symposium on Motivation* (Vol. 26, pp. 1–80). Lincoln: University of Nebraska Press.

Averill, J. R. (1982). *Anger and aggression.* New York: Springer-Verlag.

Baron, R. A., & Neuman, J. L. (1996). Workplace violence and workplace aggression: Evidence ontheir relative frequency and potential causes. *Aggressive Behavior, 22,* 161–173.

Baron, R. A., & Richardson, D. R. (1994). *Human aggression* (2nd ed.). New York: Plenum Press.

Betancourt, H., & Blair, I. (1992). A cognition (attribution)-emotion model of violence in conflictsituations. *Personality and Social Psychology Bulletin, 18,* 343–350.

Campbell, C. R., & Martinko, M. J. (1998). An integrative attributional perspective of empowerment and learned helplessness: a multimethod field study. *Journal of Management, 24*(2), 173–200.

Dodge, K. A. (1987). Hostile attribution bias among aggressive boys are exacerbated underconditions of threats to the self. *Child Development, 57*(1), 213–224.

Dodge, K. A., & Coie, J. D. (1987). Social information processing factors in reactive and proactive aggression in children's peer groups. *Journal of Personality and Social Psychology, 53,* 1146–1158.

Dodge, K. A., Murphy, R. M., & Buchsbaum, K. (1984). The assessment of intention-cue detection skills in children: Implications for developmental psychopathology. *Child Development, 55,* 163–173.

Douglas, S., & Martinko, M. (2001). Exploring the role of individual differences in the prediction of workplace aggression. *Journal of Applied Psychology. Journal of Applied Psychology, 86*(4), 547–559.

Elliott, R. H., & Jarrett, D. T. (1994). Violence in the workplace: The role of human resource management. *Public Personnel Management, 23,* 287–299.

Filipczak, B. (1993, July). Armed and dangerous at work. *Training,* pp. 39–43.

Folger, R., & Skarlicki, D. P. (1998). A popcorn metaphor for employee aggression. In R.W. Griffin, A. O'Leary-Kelly, & J. M. Collins (Eds.),*Dysfunctional behavior in organizations: Violent and deviant behavior* (pp. 43–82). Stamford, CT: JAI Press.

Gardner, W. L., & Martinko, M. J. (1988). Impression management in organizations. *Journal of Management, 14*(2), 321–338.

Greenberg, J. (1990). Employee theft as a reaction to underpayment inequity: The hidden cost of pay cut. *Journal of Applied Psychology, 75*(5), 561–568.

Grimsley, K. D. (1998, January 15). Postal peace in our time? Management has programs to defuse tensions, but labor calls for more. *Washington Post,* Financial Section, p. C1.

Kent, R., & Martinko, M.J. (1995a). The development and evaluation of a scale to measure organizational attribution style In M. J. Martinko (Ed.), *Attribution theory: An organizational perspective* (pp. 53–75). Delray Beach, FL: St. Lucie Press.

Kent, R., & Martinko, M. J. (1995b). The measurement of attributions in organizational research. In M. J. Martinko (Ed.), *Attribution theory: An organizational perspective* (pp. 17–34) Delray Beach, FL: St. Lucie Press.

Kets de Vries, M. (1993). *Leaders, fools, and imposters: Essays on the psychology of leadership.* San Francisco: Jossey-Bass.

Liden, R. C., & Mitchell, T. R. (1985). Reactions to feedback: The role of attributions. *Academy of Management Journal, 28*(2), 291–308.

Maccoby, M. (2003). *The productive narcissist: The promise and peril of visionary leadership.* Broadway Books.

Martinko, M. J., & Zellars, K. L. (1998). Toward a theory of workplace violence: A cognitive appraisal perspective. In R. W. Griffin, R. W., A. O'Leary-Kelly, & J. M. Collins (Eds.), *Dysfunctional behavior in organizations: Violent and deviant behavior* (pp. 1–42). Stamford, CT: JAI Press.

Nasby, W., Hayden, B., & DePaulo, B. M. (1979). Attributional bias among aggressive boys to interpret unambiguous social stimuli as displays of hostility. *Journal of Abnormal Psychology, 89,* 459–468.

Neuman, J. H., & Baron, R. A. (1998). Workplace violence and workplace aggression: Evidence concerning specific forms, potential causes, and preferred targets. *Journal of Management, 4*(3), 391–419.

Nunnally, J. C., Jr. (1970). *Introduction to psychological measurement.* New York: McGraw-Hill.

O'Leary-Kelly, A. M., Griffin, R. W., & Glew, D. J. (1996). Organization-motivated aggression: A research framework. *Academy of Management Review, 21*, 225–253.

Robinson, S. L., & Greenberg, J. (1998). Employees behaving badly: Dimensions, determinants, and dilemmas in the study of workplace deviance. In C. L. Cooper & D. M. Rousseau (Eds.), *Trends in organizational behavior* (Vol. 5, pp. 1–30). New York: Wiley.

Robinson, S. L., & O'Leary-Kelly, A. M. (1998). Monkey see, monkey do: The influence of work groups on the antisocial behavior of employees. *Academy of Management Journal, 41*(6), 658–672.

Russell, D. W., McAuley, E., & Tarico, V. (1987). Measuring causal attributions for success and failure: A comparison of methodologies for assessing causal dimensions. *Journal of Personality and Social Psychology, 52*(6), 1248–1257.

Skarlicki. D. P., & Folger, R. (1997). Retaliation in the workplace: The roles of distributive, procedural, and interactional justice. *Journal of Applied Psychology, 82*(3), 434–443.

Silverstein, S. (1994, March 18). The war on workplace violence. *Los Angeles Times*, National Section, p. A1.

Spector, P. E. (1982). Behavior in organizations as a function of employees' locus of control. *Psychological Bulletin, 91*, 482–497.

Storms, P. L., & Spector, P. E. (1987). Relationships of organizational frustration with reported behavioral reactions: The moderating effects of locus of control. *Journal of Occupational Psychology, 60*, 635–637.

Stuart, P. (1992). Murder on the job. *Personnel Journal, 2*, 72–77.

Toufexis, A. (1994, April 25). Workers who fight firing with fire. *Time*, pp. 35–37.

VanOostrum, N. (1997). The effects of hostile attribution on adolescents' aggressive responses to social situations. *Canadian Journal of School Psychology, 13*(1), 48–59.

Weiner, B. (1982). The emotional consequences of causal ascriptions. In N. S. Clark & S. T. Fiske (Eds.), *Affect and cognition; The 17th Annual Carnegie Symposium on Cognition* (pp. 185–200).

Weiner, B. (1985). An attributional theory of achievement motivation and emotion. *Psychology Review, 92*, 548–573.

Weiner, B. (1995). *Judgments of responsibility: A foundation of a theory of social conduct.* New York: Guilford Press.

Weiner, B., Graham, S., & Chandler, C. (1982). Pity, anger, and guilt: An attributional analysis. *Personality and Social Psychology Bulletin, 8*, 226–232.

Yates, R. E. (1995, October 1). Shining surface of U.S. business hides rising fury: Layoffs, lagging pay hikes stoke workers' anger. *Chicago Tribune*, Business Section, p. 1.

Zimbardo, P. G. (1985). *Psychology and life.* Glenview, IL: Scott Foresman.

CHAPTER 8

A PRELIMINARY EXAMINATION OF THE ROLE OF ATTRIBUTIONS AND EMOTIONS IN THE TRANSACTIONAL STRESS MODEL

An Examination of Work–Family Conflict

Kelly L. Zellars
University of North Carolina–Charlotte

**Pamela L. Perrewé, Gerald R. Ferris,
and Wayne A. Hochwarter**
Florida State University

ABSTRACT

Merging findings from the attributional (Weiner, 1985) and stress and coping literatures (Lazarus, 1966, 1993; Lazarus & Folkman, 1987), Perrewé and

Attribution Theory in the Organizational Sciences, pages 151–172
Copyright © 2004 by Information Age Publishing
All rights of reproduction in any form reserved.

Zellars (1999) expanded Lazarus' transactional appraisal approach to include a specific discussion of the process by which employees' attributions regarding stressors and the resulting emotions significantly influence their choices of coping mechanisms. Focusing on work–family conflict, this exploratory study examines the expanded transactional model in a field setting. These results provide preliminary insight into the relations among attributions, emotions, and problem solving and emotion-focused coping when experiencing work–family conflict. The implications of these results, and directions for future research, are discussed.

In his seminal work discussing the cognition-emotion process, Weiner (1985, 1986) proposed an attributional view of the emotion process, linking specific emotions to the attributions made by individuals for an event. Weiner's framework proposes that outcomes experienced by an individual trigger a general positive or negative emotion based on the perceived success or failure of the outcome. Such outcome-dependent emotions include happiness, sadness, and frustration. Following this initial emotional experience, individuals seek causes for their outcomes, thereby generating a different set of emotions dependent on the attribution made for the outcome. Such attribution-dependent emotions are determined by the perceived cause. For example, individuals might experience surprise when the perceived cause of the event is good luck, they might experience pride if the perceived cause is personal ability or effort, gratitude or anger when the perceived cause is controllability by others, or guilt given the perceived cause of controllability by self. Nearly two decades of research have generated numerous studies (e.g., Brown & Weiner, 1984, 1985; Weiner, Russell, & Lerman, 1979, Wicker, Payne, & Morgan, 1983), primarily in an achievement context, with students that provide significant evidence to support the links between attributions and emotions. Surprisingly, research efforts in this area, conducted in organizational settings, have been limited.

Nevertheless, several theoretical models linking attributions, emotions, and behaviors within the workplace have emerged. Martinko and Zellars (1998) developed an extensive model examining attributions and workplace violence. Perrewé and Zellars (1999) offered a model highlighting the importance of cognitions, attributions, and emotions arising from experienced stress in the workplace, and Moore (2000) proposed a casual attribution model for emotional exhaustion, a key component of job burnout. LePine and Van Dyne (2001) utilized the attribution–emotion model to explore the reasons peers offer to help or refrain from helping coworkers, and recently, Gundlach, Douglas, and Martinko (2003) explored the role of attributions of responsibility for whistleblowing in the workplace. Clearly, there continues to be significant interest in exploring the role of attributions and emotions in organizational settings.

Nevertheless, empirical studies examining attributions and emotions in organizational settings remain scarce. The purpose of the present study was to empirically test the relationships between attributions for work-related stress and emotions and coping choices in a field setting. First, we briefly discuss the emerging literature on emotions in the workplace, and the transactional approach to the stress process, an appraisal approach firmly rooted in the stress literature. Then, we report the results of an empirical study in which we examine the attributions made for work–family conflict, and the resulting emotions and coping choices.

EMOTIONS IN THE WORKPLACE

Because of the importance of emotions in the organizational stress process, as well as in other organizational behavior processes, it is surprising that, for decades, studies directly examining work issues and emotion remain few in number. However, beginning in the late 1980s and continuing into the 1990s, momentum for more research (e.g., Ashforth & Humphrey, 1995; Pekrun & Frese, 1992; Rafaeli & Sutton, 1987, 1989; Van Maanen & Kunda, 1989; Vecchio, 1995; Weiss & Cropanzano, 1996) in the area of emotions in organizations emerged. The underlying premise of all these studies is that "the emotional dimension is an inseparable part of organizational life and can no longer be ignored" by organizational researchers (Ashkanasy, Hartel, & Zerbe, 2000, p. 4). Included in this research is a call for more studies that examine the antecedents to frustration, guilt and shame, which remain underresearched (Roseman, Weist, & Swartz, 1994). Furthermore, Poulson (2000) recently explored some of the gaps in this area with his exploratory discussion of sources of shame in the workplace. Together, these studies indicate an expanding interest in the role of emotions in the workplace.

LAZARUS'S TRANSACTIONAL MODEL OF STRESS

A fundamental proposition of the transactional model (Folkman & Lazarus, 1984; Lazarus, 1968) is that it is the interaction of the person and environment that creates a felt stress for the individual. Emotions arising out of this interaction depend on cognitive appraisals of the significance of the person–environment relationship and available options for coping (Folkman & Lazarus, 1988). Two appraisals (i.e., primary and secondary) are central to Lazarus's cognitive theory of stress. In the primary appraisal, individuals determine if they have a stake in the outcome or how the transaction facilitates or hinders their goals (Smith & Lazarus, 1990). A stake in an encounter generates the potential for emotion (Lazarus, 1991), and stressful situations are appraised as involving harm/loss, threat, or challenge to individuals' well being (Lazarus, 1994). If individuals determine that the encounter is relevant to their well-being, a secondary appraisal occurs in which the focus is on the available coping options for altering the perceived harm, threat, or challenge, thereby creating the possibility of a more supportive or positive environment.

A significant amount of research (e.g., Folkman, 1984; Lazarus, 1966, 1968; Lazarus & Folkman, 1987) has supported the transactional model by demonstrating that the way people evaluate what is happening with respect to their well-being, and the way they cope with it, influences the occurrence and intensity of stressful experiences (Lazarus, 1993). Nevertheless, several issues remain unresolved, including the tendency of individuals to prefer certain styles of coping, and variability in findings regarding the consistency of an individual's choice of coping mechanisms (Folkman & Lazarus, 1980, 1985; MacNair & Elliott, 1992; Stone & Neale, 1984). Perrewé and Zellars (1999) proposed that one source contributing to the unresolved questions is the failure to consider the attributions and emotions arising from causal analyses of the felt stress. They relied on a long history of findings in attribution research and proposed that emotions arising from causal analysis of the felt stress by the individuals will influence their choice of coping mechanisms. An adaptation from their model is shown in Figure 8.1.

THE EXPANDED TRANSACTIONAL MODEL

Perrewé and Zellars (1999) offered a model expanding the transactional approach that highlights the importance of the cognitive and emotional components within the organizational stress process. In doing so, the authors elaborate on Lazarus's model (see Lazarus & Folkman, 1987) by explicitly including the perceived causal attributions and resulting emotions as mediating variables between the primary appraisal of felt stress and the secondary appraisal of coping choices.

Primary Appraisals and Causal Attributions

"Because of different goals and beliefs, because there is often too much to attend to, and because the stimulus array is often ambiguous, people are selective both in what they pay attention to and in what their appraisals take into account" (Lazarus, 1993, p. 9). Therefore, not all potential stressors actually cause experienced stress for an individual, and what one

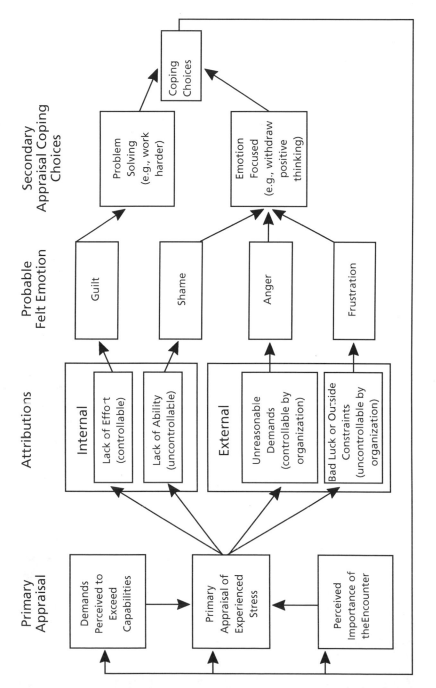

Figure 8.1. Adapted from the transacational attributional model of the organizational stress process (Perrewé & Zellars, 1999).

individual appraises as a stressful situation may not be for another. However, when the primary appraisal does reveal the presence of stressors, based on Weiner's (1985) findings, the model proposes that the employee will search for the causes of the felt stress as means to better control the environment.

This proposition appears reasonable because most employees will feel they do have a "stake in the situation" in that they desire to keep their jobs and find satisfaction in them. Although some individuals, in some demanding situations, may perceive those situations to be opportunities for achievement and experience positive emotions such as pride or positive self-esteem (Weiner, 1985), Perrewé and Zellars (1999) focused only on reactions by individuals to demanding situations associated with negative feelings such as anxiety or tension; that is, experienced negative stress.

An examination of the possible sources of stress clearly indicates that those sources can be divided along Weiner's (1985) locus of causality dimension; that is, internal causes versus external causes. For example, employees may perceive their experienced stress as an outcome arising from either a lack of effort or a lack of ability (i.e., both internal attributions). Students have been shown to use self-blame (e.g., "I realized I brought the problem on myself") when they attributed their unsatisfactory exam performance to controllable internal factors such as effort, rather than external factors such as task difficulty (Folkman & Lazarus, 1985). At the same time, other employees may perceive the stressor as being imposed upon them by external sources (e.g., the manager, the organization, or the time frame). Therefore, the causes of their stress might be the unreasonable demands of others or the difficulty of the task. The different attributions made by different employees for falling behind likely will lead to different emotional responses (Weiner, 1985). Consistent with Weiner's model (1985), potential sources of experienced stress in the expanded transactional model (Perrewé & Zellars, 1999) perceived to be internal are *effort* and *ability*. In organizational contexts, examples of external causes are *unreasonable demands* and *lack of resources*.

Specific Affective Responses to Causal Dimensions

In the transactional attribution model, employees' determination of the source of their felt stress includes an assessment of the causal dimensions and each dimension is specifically related to a set of emotions. These emotions arise from how an event is construed (Weiner, 1985), and each emotion serves a particular set of adaptive functions based on cognitive appraisals of circumstances (Smith & Ellsworth, 1985). Perrewé and Zellars

(1999) included four possible negative emotions arising from individuals' causal analysis of a stressor, that is, guilt, shame, frustration, and anger.

Consider a situation in which employees determine that the experienced stress arising from their frequent conflicts between work and family obligations is due to a lack of ability (i.e., a stable, uncontrollable, internal attribution). The researchers propose that the emotional response to such an attribution is likely to be shame, an emotion that can be described as feeling self-conscious (Roseman et al., 1994). Shame reflects "a failure or flaw of the self" (Poulson, 2000, p. 252). Alternatively, if employees perceive that the conflicts are the result of a lack of effort (i.e., an internal, unstable, controllable attribution), the authors propose that the likely response is a sense of guilt for failure to fulfill an obligation. Guilt arises when an individual's action or failure to act violates some internal standard and motivates them to repair the action (Tangney, 1992; Tangney & Fisher, 1995). Attribution theorists have found that perceptions of failure, arising from a lack of effort, have been linked to guilt and regret that may instigate motivational behaviors (Brown & Weiner, 1984; Wicker et al., 1983).

Alternatively, attribution theorists have found that the attributional antecedent for anger is an ascription of a negative, self-related outcome or event to factors that are controllable by others (Weiner, 1980a, 1980b, 1985; Weiner, Graham, & Chandler, 1982; Weiner et al., 1979). Specifically, inferences of responsibility for a negative outcome, or the breaking of a social contract, triggers anger by another party (Averill, 1982, 1983; Weiner, 1986). Anger arises from a value judgment that another person "could and should have done otherwise" (Weiner, 1995, p. 17). Based on this body of research, Perrewé and Zellars (1999) proposed that the emotional response of employees attributing their felt stress to controllable actions of the organization (e.g., unreasonable demands) will be anger.

Some studies have shown that anger is reduced to the extent information emerges that suggests the cause is uncontrollable or its outcome is unintentional or unforeseeable (Smith, Haynes, Lazarus, & Poper, 1993; Weiner, 1979, 1995). In an organization, empirical findings indicate that most employee frustration arises from a perception of interference with individuals' ability to carry out their day-to-day duties effectively (Keenan & Newton, 1984). Perrewé and Zellars (1999) proposed that the emotional response of employees attributing the source of felt stress to the situation (i.e., external and uncontrollable by either the organization or environment) is frustration.

Attributions, Secondary Appraisals, and Coping Behaviors

Secondary appraisals are evaluative processes that constantly change, and which reflect the cognitive and behavioral attempts that individuals make to manage specific internal and/or external demands appraised as taxing or exceeding personal resources (Lazarus & Folkman, 1984). Generally, in stress research, coping has two widely recognized functions: regulating stressful emotions (i.e., emotion-focused coping) and altering the troubled person–environment relationship causing the distress (i.e., problem-solving coping) (Folkman, Lazarus, Dunkeo-Schetter, DeLongis, & Gruen, 1986). In the transactional model, the coping choice considers whether the individual can act to alter a negative situation (Folkman & Lazarus, 1980, 1985; Folkman et al., 1986).

Problem-focused coping is more likely to materialize in situations that are appraised as changeable and emotion-focused coping is more likely seen in situations appraised as unchangeable (Folkman & Lazarus, 1980). Ultimately, individuals' choice of coping mechanisms are determined by their perceptions of personal control over the stressful situation, and coping outcomes, at least partially, depend on the goodness of fit between appraisal and coping (Lazarus & Folkman, 1984). Seeking information about what needs to be done and changing either individual behavior (e.g., exhibiting greater effort to keep up with the workload), or taking action on the environment, are examples of problem-solving coping efforts (Folkman et al., 1986). Based on previous findings (e.g., Smith & Ellsworth, 1985), Perrewé and Zellars (1999) proposed that employees who experience guilt, due to perceptions of a lack of effort, will engage in problem-solving efforts (e.g., work harder) as means to alter a negative situation. Thus, they proposed that guilt will mediate the relationship between individuals' perceived lack of effort and problem-solving coping.

When individuals determine that they have no means to change the situation (e.g., "These demands are unreasonable and there is nothing I can do about it"), or that they have insufficient resources (e.g., lack of ability, equipment needed), and acceptance of the stressful conditions is the only option, emotion-focused coping predominates (Folkman et al., 1986; Folkman & Lazarus, 1980; Lazarus, 1994). Emotion-focused coping efforts include distancing and escape/avoidance of the stressor, and emphasizing the positive (Folkman & Lazarus, 1985). Such efforts allow individuals to avoid focusing on the troubling situation (Folkman et al., 1986). Whereas problem-solving efforts attempt to alter the situation in a positive way, emotion-focused coping alters only the way individuals interpret situations.

Based on previous findings in stress and attributional research (e.g., Folkman & Lazarus, 1985; Weiner, 1985), Perrewé and Zellars (1999) proposed that employees who feel shame, anger, or frustration (i.e., they attribute their stress to an external source or lack of ability) will utilize emotion-focused coping as a means of reappraising a personally uncontrollable situation. Thus, shame, anger, and frustration all may be mediating variables in the attribution–secondary appraisal coping relationship. In summary, based on Perrewé and Zellars (1999), the current study tests the following hypotheses:

Hypothesis 1: *Employees who attribute the cause of felt stress to an uncontrollable, internal source (e.g., lack of ability) will report feeling shame.*

Hypothesis 2: *Employees who attribute the cause of felt stress to a controllable, internal source (e.g., lack of effort) will report feeling guilt.*

Hypothesis 3: *Employees who attribute the cause of felt stress to a controllable, external source (e.g., others in the organization) will report feeling anger.*

Hypothesis 4: *Employees who attribute the cause of felt stress to an uncontrollable, external source (e.g., the economy) will report feeling frustration.*

Hypothesis 5: *Employees who experience guilt (i.e., due to a perceived lack of effort on their part) will engage in problem-solving efforts as a means to alleviate the negative feelings.*

Hypothesis 6: *Employees who experience shame (i.e., due to a perceived lack of ability) will engage in emotion-focused efforts as a means to alleviate the negative feelings.*

Hypothesis 7: *Employees who experience anger (i.e., due to a perceived controllable but external source of demands, e.g., others in the organization) will engage in emotion-focused efforts as a means to alleviate the negative feelings.*

Hypothesis 8: *Employees who experience frustration (i.e., due to a perceived external but uncontrollable source of demands) will engage in emotion-focused efforts as a means to alleviate the negative feelings.*

WORK–FAMILY CONFLICT

We tested the hypotheses using one prevalent stressor, work–family conflict, and we chose this stressor for several reasons. First, because of the growing number of dual-career families, it is not surprising that the number of empirical (e.g., Frone, Russell, & Cooper, 1992) and qualitative (e.g., Pratt & Rosa, 2003) studies devoted to the examination of work–family conflict increased during the past decade. For example, studies have reported evidence indicating both psychological and physical costs associated with involvement in multiple roles (Frone et al., 1992), that negative moods spill over from the family environment to work and from the work environment to the family (Williams & Alliger, 1994), and that increased burnout may be a direct consequence of work–family conflict (Bacharach, Bamberger, & Conley, 1991).

Work–family conflict continues to be linked to employee job satisfaction, commitment, and intentions to withdraw from their careers (Allen, 2001; Greenhaus, Parasuraman, & Collins, 2001). Due to the prevalent experience of work–family conflict and the costs associated with this stressor, we felt it was an appropriate stressor to use in testing relationships among attributions, emotions, and coping in an organizational setting.

METHOD

Sample and Data Collection

The sample for this study was a group of professional female lawyers. A pilot study conducted with a random sample of the members confirmed that work–family conflict was a source of job stress frequently encountered by the members of the group. Anonymous questionnaires were mailed to all members of the group along with a cover letter and a postage-paid pre-addressed envelope. Three hundred ninety-one questionnaires were mailed. The questionnaires asked the individual to consider if work–family conflict was relevant to the way she felt about her job. Respondents were given the option to indicate if work–family conflict was not a stressor for them. Only respondents who reported that work–family conflict was a stressor were used in the analysis. If the stressor was relevant, respondents then rated a variety of perceived causes of the stressor, as well as the emotions that arose when the stressor was encountered on the job, and the coping mechanisms utilized by the members to cope with experienced stress. A total of 155 questionnaires were returned by mail directly to the researchers, reflecting a response rate of nearly 40% (39.6%).

Ages of respondents ranged from 21 to 83 years with a mean of approximately 40 years. Three-quarters (75%) of the members were married, 15% were single, and 7% were divorced. Forty-six percent had at least one child living at home and 17% had children not living at home.

Measures

Attributions. We measured two internal attributions and two external attributions. Both controllable and uncontrollable causes were included in the list of possible causes. Each attribution was measured by one item except for the external, controllable attribution, which was measured by two items. We divided external controllable into two dimensions; that is, attributions to others and attributions to policies. Respondents indicated the extent to which they agreed that the individual items reflected the primary cause(s) of the stressor, using a five-point scale from "strongly agree" to "strongly disagree." Higher scores indicated stronger agreement that a specific attribution reflected the cause of the stressor. An example of an item reflecting an internal controllable cause was "Me—I need to try harder." An example of an uncontrollable external cause was "No one— Stressor is caused by such things as the economy, bad luck, or other uncontrollable situation."

Emotions. Four emotions were measured: guilt, shame, anger, and frustration. Using a five-point scale ranging from "strongly agree" to "strongly disagree," respondents indicated the extent to which they agreed that each emotion reflected their feelings when the stressor was experienced. Higher scores indicate stronger agreement that the particular emotion was felt. Respondents also were given a space to indicate other emotions they felt when this stressor was experienced. Each emotion was measured using one item.

Coping. Using a five-point scale ranging from "strongly agree" to "strongly disagree," respondents indicated how they typically react to work–family conflict by indicating the extent to which they agreed that the individual items reflected their reactions to the stressor. Higher scores indicated stronger agreement that they behave in the manner indicated by the item. Emotion-focused coping was measured with two items adapted from Beehr, King and King (1990): "I focus on the good things in my life away from work" and "I withdraw or distance myself from the situation." The two items yielded a Cronbach alpha of .90. Problem-solving coping was measured with five items adapted from Beehr and colleagues. Examples include "I break the problem down into steps" and "I ask for help to get the job done." The five items yielded a Cronbach alpha of .96.

Control variables. Based on previous research regarding correlates of work–family conflict (Kossek & Ozeki, 1998), we controlled for marital status, number of children living at home, and age of the respondent. For marital status, respondents checked a box for married and living with spouse (1), single (2), divorced (3), and other (4). Respondents were given four options to indicate the number of children living at home. The options were "none," "1 or 2," "3 or 4," and "more than 4" (4). Age was recorded by entering their age in years.

RESULTS

Table 8.1 presents the means, standard deviations, and intercorrelations of the variables used in the analysis. Due to missing observations, and because only those respondents who indicated work–family conflict was a stressor for them were used, the sample size was reduced to 111 respondents.

Table 8.1. Means, Standard Deviations, and Intercorrelations of Variables

Variables	M	sd	1	2	3	4	5	6	7	8	9	10	11	12	13	14
1. Age	39.6	9.35	1.00													
2. Marital status	1.36	.70	.17*	1.00												
3. Children	1.5	.59	-.33**	-.33	1.00											
4. Internal (effort)	3.0	1.38	-.04	-.12	-.04	1.00										
5. Internal (ability)	2.07	1.14	-.02	.12	-.06	.52**	1.00									
6. External (others)	3.5	1.13	-.11	-.14	.02	.13	.18	1.00								
7. External (policies)	2.86	1.30	-.19	.05	.09	-.02	.16	.47**	1.00							
8. External (no one)	3.21	1.32	-.09	-.08	.10	-.06	-.07	.00	.04	1.00						
9. Shame	2.26	1.36	-.04	.06	.02	.40**	.62**	.26**	.31**	-.02	1.00					
10. Guilt	3.73	1.27	.13	-.10	.36**	.21*	.22*	.21*	.21*	-.04	.32**	1.00				
11. Frustration	4.17	.88	-.19*	-.26**	.21*	.23*	.09	.31**	.17	.19*	.08	.20*	1.00			
12. Anger	3.23	1.37	-.06	.16	-.02	.32**	.46**	.22*	.44**	.08	.53**	.31**	.32**	1.00		
13. Emotion-focused coping	3.17	.76	-.04	-.05	-.01	.37**	.26**	.22*	.19*	.10	.42**	.08	.09	.39**	1.00	
14. Problem-solving coping	3.34	.68	.19*	-.02	.18	.31**	.21*	.23*	.11	.11	.23*	.37**	.10	.10	.28**	1.00

** $p < .01$; * $p < .05$

An examination of the correlation matrix provides support for most of the hypothesized relationships. All proposed relations between specific attributions and corresponding emotions were supported. Specifically, attribution to a lack of effort were correlated with feelings of guilt; attributions to a lack of ability were correlated with shame; attributions to others and organizational policies (both external and controllable) were correlated with anger; and attributions to no one as the cause of the stressor was correlated with frustration. However, several of the attributions correlated with other emotions besides the hypothesized one. In order to provide a more conservative test of these relationships and to control for some of the multicolinearity, hierarchical regression was utilized to test the hypothesized relationships. The models predicting that attributions would affect emotions were examined first. For each emotion (i.e., guilt, shame, frustration, anger), the demographic variables (i.e., marital status, age, and children) were entered first followed by the main effect variables (i.e., internal and external attributions). The models for shame, anger, guilt, and frustration were not statistically significant. In both cases, having children was the only significant predictor. Therefore, none of the hypotheses focusing on the impact of attributions on emotions (Hypotheses 1–4) was supported.

To test the coping hypotheses (Hypotheses 5–8), the demographic variables were entered followed by the attributions and emotions. Table 8.2 indicates the results of the analyses regarding problem–solving efforts. Age was significantly and positively associated with problem-solving coping in the first step. Perceiving a lack of effort (i.e., internal, controllable attribution) was also positively associated with the use of problem-solving efforts. Furthermore, after controlling for the attributions, the emotion of guilt was also positively associated with the use of problem-solving efforts, providing some support for Hypothesis 5. The overall model was significant ($F = 2.73$, $p < .001$) and explained 25% of the variance in the use of problem-solving efforts among respondents.

Table 8.2. Hierarchical Regression Results Predicting Problem-Solving Coping

Dependent Variable: Problem-Solving	β	se	β	se	β	se
Step 1: Control Variables						
Marital Status	–.07	.12	–.01	.12	.00	.00
Child	.14	.10	.16	.10	.14	.15
Age	.27**	.01	.25	.01	.21*	.02
Step 2: Attributions						
Effort toward job			.26*	.05	.27*	.05

Table 8.2.   Hierarchical Regression Results Predicting Problem-Solving Coping (Cont.)

Dependent Variable: Problem-Solving	β	se	β	se	β	se
Ability to do job			.01	.05	−.03	.05
Other people in/out of organization			.19	.05	.12	.05
Organizational policies or regulations			−.05	.04	−.03	.05
No one—economy, bad luck, or other uncontrollable situation			.02	.04	.04	.04
Step 3: Emotions						
Shame					.03	.04
Guilt					.24*	.05
Frustration					−.11	.07
Anger					−.10	.05

Regarding the use of emotion-focused coping efforts, Table 8.3 presents the results of the analysis. The overall model was significant ($F = 2.41$, $p < .01$) and explained 23% of the variance in the use of emotion coping. None of the control variables was significant. Perceiving the cause to be a lack of effort was positively associated with the use of emotion coping. Interestingly, perceiving the cause of work–family conflict to be the ability to do the job was negatively associated with the use of emotion coping. Controlling for attributions and consistent with Hypothesis 6, respondents' feelings of a greater sense of shame were positively associated with using more emotion-focused efforts to react to the situation, and feelings of guilt were negatively associated with using more emotion-focused coping. Hypotheses 7 and 8 were not supported as neither feelings of anger or frustration were associated with the use of emotion-focused coping. However, an examination of the correlation matrix provides support for three of the four hypothesized relationships. Those experiencing guilt reported using problem-solving coping techniques; and, as expected, there was no relationship between guilt and emotion-focusing coping. Those lawyers experiencing shame reported using emotion-focused coping as well as problem-solving coping. When respondents experienced anger, they reported using emotion-focused coping; as expected, no relationship was found for problem-solving coping. Contrary to the hypothesis, there were no significant correlations for those reporting feelings of frustration and coping mechanisms.

Table 8.3.   Hierarchical Regression Results Predicting Emotion-Focused Coping

Dependent Variable: Emotion Focused	β	se	β	se	β	se
Step 1: Control Variables						
Marital Status	−.13	.14	−.11	.15	−.14	.14
Child	−.09	.12	−.08	.12	−.01	.12
Step 2: Attributions						
Effort toward job			.29*	.06	.28*	.06
Ability to do job			−.15*	.06	−.23*	.06
Other people in/out of organization			−.07	.06	−.07	.05
Organizational policies or regulations			.08	.05	.04	.05
No one—economy, bad luck, or other uncontrollable situation			−.03	.04	−.03	.04
Age	.06	.01	.05	.01	.07	.01
Step 2: Attributions						
Effort toward job			.29*	.06	.28*	.06
Ability to do job			−.15*	.06	−.23*	.06
Other people in/out of organization			−.07	.06	−.07	.05
Organizational policies or regulations			.08	.05	.04	.05
No one—economy, bad luck, or other uncontrollable situation			−.03	.04	−.03	.04
Step 3: Emotions						
Shame					.35*	.05
Guilt					−.21	.05
Frustration					−.04	.07
Anger					.22	.06

DISCUSSION

Contributions of the Study

These results provide a preliminary insight into the relations among attributions, emotion, and coping when experiencing work–family conflict.

With regard to coping with work–family conflict, results indicated that age of the respondent, a lack of effort, and feelings of guilt were associated positively with problem-solving coping. As expected, the attribution of lack of effort and the emotion of feeling shame were associated positively with emotion-focusing coping. Perhaps these individuals reflect a type of learned helplessness (Seligman, 1974), reflecting more passive or acquiescent behaviors. Learned helplessness is rooted in the belief that regardless of a person's actions, unpleasant outcomes are unavoidable. Individuals experiencing learned helplessness perceive that despite their best effort to change a situation, unpleasant outcomes cannot be avoided (Overmier & Seligman, 1967; Seligman, 1974). Interestingly, attributions made to ability were negatively related to emotion-focused coping, indicating that when individuals make an attribution that is uncontrollable and internal, emotion-focused coping is less likely to occur. The associations among attributions, emotions, and each coping choice (i.e., problem solving and emotion focused) are examined in more detail.

Problem-solving coping. Individuals who report their experienced stress from work–family conflict is due to a lack of effort (i.e., internal, controllable attribution) are more likely to use problem-solving coping mechanisms. Because these individuals are attributing their strain to their own personal failure, attribution theorists have argued that this may motivate them to make a behavioral change (e.g., Brown & Weiner, 1984). Furthermore, feelings of guilt, the only emotion expected to be associated with the use of problem-solving mechanisms, are associated positively with problem-solving coping. Because guilt occurs when individuals perceive failure due to their own actions, they are motivated to remedy or correct the action (e.g., Tangney & Fisher, 1995). The motivation to remedy or correct an action can be seen as a form of problem-solving coping. In contrast to learned helplessness, such individuals reflect reactance theory in that they are motivated to engage in behaviors that reestablish the free behaviors, which have been eliminated (e.g., spending sufficient time with children) or are threatened with elimination (Brehm, 1966). Problem solving may reflect an attempt to regain lost control, a form of reactance.

Finally, although age was examined as a control variable, it is interesting to note that older employees are more likely to report using problem-solving coping mechanisms. Perhaps as individuals get older and gain experience, they are more confident in their abilities to change a stressful situation, or perhaps they are better able to see opportunities to change a situation.

Emotion-focused coping. Individuals attributing their strain to internal, controllable causes (i.e., lack of effort), are more likely to report using emotion-focused coping. Interestingly, those attributing their strain to internal, uncontrollable causes (i.e., lack of ability) are less likely to report using emotion-focused coping. Neither of these relationships was predicted. In regard to making attributions to a lack of effort, these individuals are attributing their strain to their own personal failure, thus, they may be searching for *any and all* coping mechanisms that might help remedy the situation. Because a lack of effort is associated with both problem-solving and emotion-focused coping, making an attribution that the cause is internal (i.e., internal, controllable) may lead individuals to utilize a variety of coping strategies.

In regard to attributions to lack of ability, perhaps the hopelessness of realizing individuals are not able to remedy the situation results in using fewer emotion-focused coping mechanisms. Specifically, these individuals may not even try to relieve the experienced strain. Because neither of these relationships was predicted, additional research is needed. As predicted, feelings of shame were associated with reporting more use of emotion-focused coping. Because the emotion of shame comes from the perception of a personal flaw (Poulson, 2000), individuals might attempt to deal with the strain through withdrawal, avoidance, or cognitive reappraisal. Trying to change the situation may seem fruitless to those experiencing this emotion. Shame was not associated with problem-solving coping, which supports the notion that internal, uncontrollable attributions are more likely to be associated with emotion-focused coping than problem-solving coping.

Limitations

Several limitations of the present study should be acknowledged. First, examining a process model using cross-sectional data, although commonly seen in the literature, does not capture the dynamics of the attributional stress phenomenon. Second, due to the nature of the study and a lack of available standardized measures at the time of the data collection, we used a number of one-item measures that may not have captured the true construct of interest, and, of course, one-item measures have unknown psychometric properties. Furthermore, the sample was entirely female; given there have been documented differences in emotional and aggressive behaviors between men and women (Eagly & Steffan, 1986; Feshbach, 1997), this may have affected the strength of the relationships. For example, after displaying aggressive behavior, women are more likely than men to experience fear, anxiety, and guilt (Eagly & Steffan, 1986).

Implications for Practice and Future Research

Momentary emotional states can have long-lasting effects on individual attitudes and behaviors (Weiss & Cropanzano, 1996). Positive emotions promote individuals' desires to maintain and prolong their current feeling state (Weiss & Cropanzano, 1996). Thus, positive emotions should be positively related to staying with and commitment to the organization (Ashkanasy & Daus, 2002). Negative emotions, on the other hand, promote individuals' efforts to avoid similar situations. One coping mechanism or option for avoidance is to leave the organization. Clearly, as most managers would like to increase organizational commitment and reduce the amount of turnover in the organization, attending to employees' attributions and resulting emotions may be a viable means to achieve this.

CONCLUSION

In summary, we found that attributions were associated with numerous emotions simultaneously, not simply one specific emotion. Thus, although attributions may affect the choice of coping mechanisms, it may be through emotions in general, but it may not be through distinct emotions. Managers need to understand that changes in the organizational environment might lead to a number of different emotions. However, it is the attributions employees make about an organizational change will likely affect the way they cope with the change.

Recent research by Rotondo, Carlson, and Kincaid (2003) has indicated that problem-solving coping behaviors are associated with fewer conflicts between family and work, whereas avoidance coping, a form of emotion-focused coping, is associated with greater work–family conflict. Because the present study found problem-solving coping to be associated with internal, controllable attributions, perhaps efforts to help individuals attribute certain types of experienced stress to themselves due to a lack of effort (i.e., internal and controllable) might eventually reduce the level of conflict experienced. This study should be considered to be an initial exploration into the complex relationships among perceived stressors, attributions, emotions, and coping. Although our empirical study was an initial step toward understanding how and why people choose certain coping mechanisms via the expanded version of Lazarus's transactional model of stress (Lazarus, 1994), additional research using a variety of different stressors is critical to the continued development of this work.

REFERENCES

Allen, T. D. (2001). Family-supportive work environments: The role of organizational perceptions. *Journal of Vocational Behavior, 58,* 414–435.

Ashforth, B. E., & Humphrey, R. H. (1995). Emotion in the work place: A reappraisal. *Human Relations, 48,* 97–125.

Ashkanasy, N. M., & Daus, C. S. (2002). Emotion in the workplace: the new challenge for managers. *Academy of Management Executive, 16,* 76–86.

Ashkanasy, N. M., Hartel, C. E. J., & Zerbe, W. J. (2000). Emotions in the workplace: Research, theory, and practice. In N. M. Ashkanasy, C. E. J. Hartel, & W. J. Zerbe (Eds.), *Emotion in the workplace* (pp. 1–18). Westport, CT: Quorum Books.

Averill, J. R. (1982). *Anger and aggression: An essay on emotion.* New York: Springer-Verlag.

Averill, J. R. (1983). Studies on anger and aggression. *American Psychologist, 38,* 1145–1160.

Bacharach, S. B., Bamberger, P., & Conley, S. (1991). Work-home conflict among nurses and engineers: Mediating the impact of role stress on burnout and satisfaction at work. *Journal of Organizational Behavior, 12,* 39–45.

Beehr, T., King, L. A., & King, D.W. (1990). Social support and talking to supervisors. *Journal of Vocational Behavior, 36,* 61 –81.

Brehm, J. (1966). *A theory of psychological reactance.* New York: Academic Press.

Brown, J., & Weiner, B. (1984). Affective consequences of ability versus effort ascriptions: controversies, resolutions, and quandaries. *Journal of Educational Psychology, 76,* 146–158.

Douglas, S. C., & Martinko, M. J. (2001). Exploring the role of individual differences in the prediction of workplace aggression. *Journal of Applied Psychology, 86,* 547–559.

Eagly, A., & Steffan, V. J. (1986). Gender and aggressive behavior: A meta-analytic review of the social psychological literature. *Psychological Bulletin, 100,* 309–330.

Feshbach, S. (1997). The psychology of aggression: Insights and issues. In S. Feshbach & J. Zagrodzka (Eds.), *Aggression: Biological, developmental, and social perspectives* (pp. 213–235). New York: Plenum Press.

Folkman, S. (1984). Personal control and stress and coping processes: A theoretical analysis. *Journal of Personality and Social Psychology, 48,* 839–852.

Folkman, S., & Lazarus, R. S. (1980). An analysis of coping in a middle-aged community sample. *Journal of Health and Social Behavior, 21,* 219–239.

Folkman, S., & Lazarus, R. S. (1985). If it changes it must be a process: Study of emotion and coping during three stages of a college examination. *Journal of Personality and Social Psychology, 48,* 150–170.

Folkman, S., & Lazarus, R. S. (1988). Coping as a mediator of emotion. *Journal of Personality and Social Psychology, 54,* 466–475.

Folkman, S., Lazarus, R. S., Dunkel-Schetter, C., DeLongis, A., & Gruen, R.J. (1986). Dynamics of a stressful encounter: Cognitive appraisal, coping, and encounter outcomes. *Journal of Personality and Social Psychology, 50 ,* 992–1003.

Frone, M. R., Russell, M., & Cooper, M. L. (1992). Antecedents and outcomes of work-family conflict: Testing a model of the work-family interface. *Journal of Applied Psychology, 77,* 65–75.

Gundlach, M. J., Douglas, S. C., & Martinko, M. J. (2003). The decision to blow the whistle: A social information processing approach. *Academy of Management Review, 28,* 107–123.

Greenhaus, J. H., Parasuraman, S., & Collins, K. M. (2001). Career involvement and family involvement as moderators of the relationship between work-family conflict and withdrawal from a profession. *Journal of Occupational Health Psychology, 6,* 91–100.

Keenan, A., & Newton, T.J. (1984). Frustration in organizations: Relationships to role stress, climate, and psychological strain. *Journal of Occupational Psychology, 57,* 57–65.

Kossek, E., & Ozeki, C. (1998). Work-family conflict, policies, and the job-life satisfaction relationship: A review and directions for organizational behavior-human resources research. *Journal of Applied Psychology, 83,*139–149.

Lazarus, R. S. (1966). *Psychological stress and the coping process.* New York: McGraw-Hill.

Lazarus, R. S. (1968). Emotions and adaptation: Conceptual and empirical relations. In W. J. Arnold (Ed.), *Nebraska Symposium on Motivation* (pp. 175–266). Lincoln: University of Nebraska Press.

Lazarus, R. S. (1982). Thoughts on the relations between emotion and cognition. *American Psychologist, 37,* 1019–1024.

Lazarus, R. S. (1991). Progress on a cognitive-motivational-relational theory of emotion. *American Psychologist, 46,* 819–834.

Lazarus, R. S. (1993). From psychological stress to the emotions: A history of changing outlooks. *Annual Review of Psychology, 44,* 1–21.

Lazarus, R. S. (1994). Psychological stress in the workplace. In R. Crandall & P. L. Perrewé (Eds.), *Occupational stress: A handbook* (pp. 1–5). New York: Taylor & Francis.

Lazarus, R. S., & Folkman, S. (1984). *Stress, appraisal, and coping.* New York: Springer.

Lazarus, R. S., & Folkman, S. (1987). Transactional theory and research on emotions and coping. *European Journal of Personality, 1,* 141–169.

LePine, J. A., & Van Dyne, L. (2001). Peer responses to low performers: An attributional model of helping in the context of groups. *Academy of Management Review, 26,* 67–84.

MacNair, R. R., & Elliott, T. R. (1992). Self-perceived problem-solving ability, stress appraisal, and coping over Time. *Journal of Research in Personality, 26,* 150–164.

Martinko, M. J., & Zellars, K. L. (1998). Toward a theory of workplace violence and aggression: A cognitive appraisal perspective. In R. W. Griffin, A. O'Leary-Kelly, & J. M. Collins (Eds.), *Dysfunctional behavior in organizations: Violent and deviant behavior* (pp. 1–42), Stamford, CT: JAI Press.

Moore, J. (2000). Why is this happening? A causal attribution approach to work exhaustion consequences. *Academy of Management Review, 25,* 335–349.

Overmier, J. B., & Seligman, M. E. P. (1967). Effects of inescapable shock upon subsequent escape and avoidance learning. *Journal of Comparative and Physiological Psychology, 63,* 28–33.

Pekrun, R., & Frese, M. (1992). Emotions in work and achievement. In C. L. Cooper & D. T. Robertson (Eds.), *International review of industrial and organizational psychology* (Vol. 7, pp. 153–200). St. Edmundsbury Press, UK: Wiley.

Perrewé, P. L., & Zellars, K. L. (1999). An examination of attributions and emotions in the transactional approach to the organizational stress process. *Journal of Organizational Behavior, 20,* 739–752.

Pratt, M. G., & Rosa, J. A. (2003). Transforming work-family conflict into commitment in network marketing organizations. *Academy of Management Journal, 46,* 395–418.

Poulson, C. F. II. (2000). Shame and work. In N. M. Ashkanasy, C. E. J., Hartel, & W. J. Zerbe (Eds.), *Emotion in the workplace* (pp. 250–271). Westport, CT: Quorum Books.

Rafaeli, A., & Sutton, R. I. (1987). Expression of emotion as part of the work role. *Academy of Management Review, 12,* 23–27.

Rafaeli, A., & Sutton, R. I. (1989). The expression of emotion in organizational life. In L. L. Cummings & B. M. Staw (Eds.), *Research in organizational behavior* (Vol. 11, pp. 1–42). Greenwich, CT: JAI Press.

Roseman, I. J., Weist, C., & Swartz, T. S. (1994). Phenomenology, behaviors, and goals differentiate discrete emotions. *Journal of Personality and Social Psychology, 67,* 206–221.

Rotondo, D. M., Carlson, D. S., & Kincaid, J. F. (2003). Coping with multiple dimensions of work-family conflict. *Personnel Review, 32,* 275–296.

Seligman, M. E. P. (1974). Depression and learned helplessness. In R. J. Friedman & M. M. Katz (Eds.), *The psychology of depression: Contemporary theory and research* (pp. 83–113). Washington, DC: Winston-Wiley.

Smith, C. A., & Ellsworth, P. C. (1985). Patterns of cognitive appraisal in emotion. *Journal of Personality and Social Psychology, 48,* 813–838.

Smith, C., Haynes, K. N., Lazarus, R. S., & Pope, L. K. (1993). In search of the "hot" cognitions: attributions, appraisals, and their relation to emotion. *Journal of Personality and Social Psychology, 65,* 916–929.

Stone, A. A., & Neale, J.M. (1984). A new measure of daily coping: Development and preliminary results. *Journal of Personality and Social Psychology, 46,* 892–906.

Tangney, J. (1992). Situational determinants of shame and guilt in young adulthood. *Personality and Social Psychology Bulletin, 18,* 199–206.

Tangney, J., & Fischer, K. (1995). Self-conscious emotions: *The psychology of shame, guilty, guilty, embarrassment, and pride.* New York: Guilford Press.

Van Maanen, J., & Kunda, G. (1989). Real feelings: Emotional expression and organizational culture. In L. L. Cummings & B. M. Staw (Eds.), *Research in organizational behavior* (Vol. 11, pp. 43–104). Greenwich, CT: JAI Press.

Vecchio, R. P. (1995). It's not easy being green: Jealousy and envy in the workplace. In G. R. Ferris (Ed.), *Research in personnel and human resources management* (Vol. 13, pp. 201–244). Greenwich, CT: JAI Press.

Weiss, H. M., & Cropanzano, R. (1996). Affective events theory: A theoretical discussion of the structure, causes, and consequences of affective events at work. In L. L. Cummings & B. M. Staw (Eds.), *Research in organizational behavior* (Vol. 18, pp. 1–50). Greenwich, CT: JAI Press.

Weiner, B. (1979). A theory of motivation for some classroom experiences. *Journal of Educational Psychology, 71,* 3–25.

Weiner, B. (1980a). A cognitive (attribution)-emotion-action model of motivational behavior: An analysis of judgments of help-giving. *Journal of Personality and Social Psychology, 39*, 186–200.

Weiner, B. (1980b). May I borrow your class notes? An attributional analysis of judgments of help-giving in an achievement related context. *Journal of Educational Psychology, 72*, 676–681.

Weiner, B. (1985). An attributional theory of achievement motivation and emotion. *Psychological Review, 92*, 548–573.

Weiner, B. (1986). *An attributional theory of achievement motivation and emotion.* New York: Springer-Verlag.

Weiner, B. (1995). *Judgments of responsibility.* New York: Guilford Press.

Weiner, B., Graham, S., & Chandler, C. (1982). Causal antecedents of pity, anger, and guilt. *Personality and Social Psychology Bulletin, 8*, 226–232.

Weiner, B., Russell, D., & Lerman, D. (1979). The cognition-emotion process in Achievement-related contexts. *Journal of Personality and Social Psychology, 37*, 1211–1220.

Weiss, H. M., & Cropanzano, R. (1996). Affective events theory: a theoretical discussion of the structure, causes and consequences of affective experiences at work. *Research in Organizational Behavior, 18*, 1–74.

Wicker, F. W., Payne, G. C., & Morgan, R. D. (1983). Participant descriptions of guilt and shame. *Motivation and Emotion, 7*, 25–39.

Williams, J. K., & Alliger, M. G. (1994). Role stressors, mood spillover, and perceptions of work-family conflict in employed parents. *Academy of Management Journal, 37*, 837–868.

Zajonc, R. B. (1984). On the primacy of affect. *American Psychologist, 39*, 117–123.

Zajonc, R. B., & Markus, H. (1985). Must all affect be mediated by cognition? *Journal of Consumer Research, 12*, 363–364.

CHAPTER 9

SOCIAL ATTRIBUTIONAL STYLE

A Conceptual and Empirical Extension of Attributional Style

Neal F. Thomson
Columbus State University

Mark J. Martinko
Florida State University

ABSTRACT

The importance of observer attributions within organizational and leadership contexts is well documented. Although numerous studies confirm that individuals have consistent patterns of self-attributions regarding their own successes and failures, the idea that observers may also have attribution styles has not received attention. This study examines the premise that individuals have cross-situational consistencies (i.e., attribution styles) that bias their evaluations of others in the workplace. It describes the theoretical rationale, construction, factor structure, and construct validation of a scale designed to measure social attribution styles in the workplace. The results support the construct of a social attribution style. Implications of social attribution styles for leadership and management, particularly with respect to appraisals and

Attribution Theory in the Organizational Sciences, pages 173–201

judgments of the causes of member and organizational performance, are discussed. Suggestions for further research are also provided.

Over the last two decades, attribution theory has emerged as an important theoretical foundation for explaining and interpreting organizational behavior (Martinko, 1995). Two streams of research have appeared. The first, following Weiner's (1985) theory of achievement motivation, has focused on the intrapersonal process of how individuals interpret organizational outcomes relating to their personal successes and failures (e.g., Adler, 1980; Bettman & Weitz, 1983; Clapham & Schwenk, 1991; Liden & Mitchell, 1985; Mone & Baker, 1992; Salancik & Meindl, 1984). The second stream of research, which is usually based on Kelley's (1973) cube, has focused on social (i.e, observer) attributions, the process by which individuals assign causation for the behavior of others. In particular, social attributions have permeated research on the interpersonal processes associated with leader–member relations such as performance appraisal (Ashkanasy, 1995; Blakely, 1993; Fedor & Rowland, 1989; Ferris, Yates, Gilmore, & Rowland, 1985; Heneman, Greenberger, & Anonyuo, 1989: Knowlton & Mitchell, 1980), disciplinary behaviors of leaders (Anderson, 1992; Dobbins & Russell, 1986; Green & Liden, 1980; Hesketh, 1984; Klaas & Wheeler, 1990; Mitchell & Kalb, 1982; Mitchell & Wood, 1980; Moss & Martinko, 1997), and perceptions of leadership effectiveness (Cronshaw & Lord, 1987).

Past work by Ashkanasey (1989, 1995); Dykman and Abramson (1990), and Martinko and Thomson (1998) has explained and demonstrated how the work on both self and social attributions can be integrated by mapping Kelley's dimensions of information map onto Weiner's dimensions and corresponding attributional explanations. Thus, for example, the synthesized model presented by Martinko and Thomson illustrates how information that is low in consensus, high in distinctiveness, and high in consistency leads to internal stable and specific attributions such as ability.

While there are many specific implications for both theory and research suggested by the integration of the Kelley and Weiner models, a more general implication, which is axiomatic to this chapter, is that both social and self attributions are reflections of the same general causal reasoning process. Thus, research on self attribution should inform research in social attribution and vice versa. In particular, one of the most important results of the research on self-attributions has been the repeated finding that people demonstrate cross situational stabilities (i.e., attribution styles) in the way they attribute causes for their outcomes (Abramson, Seligman, & Teasdale, 1978; Cutrona, Russell, & Jones, 1985; Peterson et al., 1982). Thus, individuals can be classified as optimistic, which is characterized as a bias toward explaining failures in terms of external, unstable, and specific causes such as chance and by explaining success in terms of internal, stable, and global causes such as ability. On the other hand, pessimistic styles are characterized by internal, stable, and global attributions for failure such as lack of ability and by attributions for success that are characterized by external, unstable, and specific attributions such as chance (Seligman, 1990). Most importantly, these styles have been shown to have an effect on both affect and behavior. Thus, for example, Seligman and Schulman (1986) have demonstrated that successful salespeople have a consistently optimistic bias and Deiner and Dweck (1978) have demonstrated that pessimistic attributions are related to both attenuated behavioral responses and depressed affect.

The basic premise for the research described in this chapter is that, assuming that the processes of social and self attributions reflect a similar underlying process, it is likely that observers exhibit the same type of cross-situational stability (i.e., attributional style) that has repeatedly been shown to be present in self-attributions. Moreover, because the underlying process of causal reasoning is likely to share some elements, we are initially proposing that the same dimensions and explanations used to characterize self attribution styles will also be characteristic of social attribution styles. This, however, is based on the assumption that the two processes rely on the same subprocesses, which is a supposition that may or may not be supported by the data.

Furthermore, given that attribution styles have been demonstrated to be associated with both affective and behavioral tendencies with regard to self attributions, we expect that social attribution styles will also be related to the behavior and affect that observers (e.g., leaders) display toward actors (e.g., members). Thus, within an organizational context, we expect that the construct of social attribution style will have broad implications for how leaders feel and behave toward members.

Based on the above rationale and support for the construct of Social Attributional Style (SAS), two studies were conducted. The first study involves the development and assessment of the psychometric properties of an instrument designed to measure SAS. The second study describes the process of validating the new measure and construct of SAS by testing the theoretical relationships that would be expected if the construct of a SAS is valid.

STUDY 1: SCALE DEVELOPMENT

Accepting the premise that the basic processes of self and social attributions are the same, it follows that the procedures for measuring SASs will be similar to the procedures for measuring self-attribution styles. Numerous scales have been developed to measure self-attribution styles (Campbell & Martinko, 1998; Cutrona, Russell, & Jones, 1985; Kent & Martinko, 1995; Peterson et al., 1982; Peterson & Villanova, 1988; Russell, 1982). Common to all of these scales are the dimensions of locus of causality (LOC) (i.e., internal/

external) and stability (stable/unstable), while other dimensions such as controllability (i.e., controllable/uncontrollable), intentionality (intentional/unintentional), and globality (i.e., global/specific) are somewhat less common. As Weiner (1985) notes, the relevant dimensions of attributions are dependent upon the domain of interest. Thus, for example, intentionality is not normally included in scales of self attribution style since people rarely intend negative outcomes for themselves. On the other hand, intentionality makes sense for SASs because observers may believe that an actor intentionally caused a negative outcome.

In the context of the current study, we reasoned that, as in the case of almost all of the self attribution style scales, we would include the dimensions of LOC and stability because they are clearly theoretically relevant with respect to social attributions (Kelley, 1973; Weiner, 1985). In addition, because our scale was exploratory, we wanted to make sure that we included all of the dimensions for SASs that are potentially relevant, since we could reject irrelevant dimensions in the refinement of our scales. Based on the works of Weiner (1985), Kelley (1973), Martinko and Thomson (1998), and Abramson and colleagues (1978), we included the dimensions of intentionality, controllability, and globality. Our general research questions are:

> Q1: Is there cross situational consistency in the dimensions of the social attributions such that a scale to measure SAS can be developed that meets the standard of reliability of .70 or higher suggested by Nunnally (1978).
>
> Q2: From among the dimensions of internality, stability, globality, controllability, and intentionality, what subset of dimensions provides the most accurate and reliable description of SAS.

METHOD: STUDY I

General Procedure

The first study was conducted in two stages. In the first stage, an SAS questionnaire was pilot tested by administering it to a group of subjects who were asked to provide feedback regarding readability and clarity. This was followed by a large-scale administration of the questionnaire to test the reliability and factor structure of the instrument.

Pilot Study

Preliminary data were collected from a group of 31 undergraduate students who were asked to complete a copy of the initial draft of a SAS questionnaire. They were asked to note any problems involving clarity of scenarios, instructions, or the questionnaire format. Minor problems that were identified during this phase were subsequently corrected on the final draft of the SAS questionnaire.

Procedures and Data Collection

The next wave of data was collected from 375 junior/senior undergraduate business students enrolled in a large public university in the southern United States. All subjects received a letter asking for their voluntary participation and were then asked to complete the SAS questionnaire and a demographic information sheet. Three hundred and two usable questionnaires were returned for a response rate of 81%.

The demographics of the sample closely reflected the makeup of the workforce, with slightly more males than females (see Table 9.1). The majority of the students reported 5 or more years of work experience. Less than 10% of the respondents had no work experience.

Table 9.1.   Sample Characteristics: Study 1

Characteristic	Number	%
Sex		
Male	210	56
Female	163	43.5
Not Reported	2	0.5
Year		
Sophomore	3	.8
Junior	94	25.1
Senior	275	73.3
Graduate	1	0.3
Not Reported	2	0.5
GPA		
2.01–2.5	37	9.9
2.51–3.0	154	41.1
3.01–3.5	131	34.9
3.51–4.0	52	13.9
Not Reported	1	0.3
Work Experience		
None	11	2.9

Table 9.1.   Sample Characteristics: Study 1 (Cont.)

Characteristic	Number	%
Less than 1 year	32	8.5
1–2 Years	47	12.5
2–3 Years	54	14.4
3–5 Years	77	20.5
More than 5 Years	153	40.8
Not Reported	1	0.3
Managerial/Supervisory Experience		
Yes	170	45.3
No	195	52
Not Sure	10	2.7
Reported Academic Major		
Management	47	12.5
Human Resource Management	24	6.4
Finance	51	13.6
Accounting	49	13.1
Marketing	54	14.4
All others (including dbl. majors)	149	39.7
Not Reported	1	0.3

The Social Attribution Style Questionnaire (SASQ)

As previously noted, there are a variety of questionnaires that measure self-attribution styles. These scales include the Attributional Style Questionnaires (ASQ) (Peterson et al., 1982; Peterson & Villanova, 1988), Organizational Attribution Style Questionnaire (OASQ) (Campbell & Martinko, 1998; Kent & Martinko, 1995), and the Causal Dimension Scale (CDS) (Russell, 1982). Since, as suggested by the Martinko and Thomson (1998) model, the dimensions of both self and social attributions are essentially the same, any of these questionnaires could be adapted to an observer perspective by simply changing the target of each question from the self to others. We chose to use the OASQ, developed by Kent and Martinko (1995), since the events were related to the work domain and the scale has been found to have relatively robust psychometric properties, particularly with respect to the reliabilities of the dimensions. Therefore, we used the OASQ as a model for developing the SASQ.

The OASQ is a Likert-type instrument that includes the dimensions of LOC, stability, controllability, and globality. There are 16 occurrences, each relating to a negative outcome, or failure, in the work environment. For each of these situations, the subject is asked to rate their attribution on each of the dimensions on a scale of one to seven. Reliabilities on this instrument are respectable, with Cronbach's alphas on one revision as high as .70 for locus of causality, .81 for stability, .72 for controllability, and .84 for globality (Thomson & Martinko, 1995)

The SASQ for this study was developed using situations similar to the OASQ, but framed in an observer's perspective. As with the OASQ, 16 situations were described, each with a negative work-related outcome. The questionnaire then asked the subjects their attributions for the actors' outcomes and asked them to classify the attributions according to the five dimensions of locus of causality, stability, globality, controllability, and intentionality. This questionnaire used the same seven-point, Likert-type scale as the OASQ (see the Appendix). The similarity between these two questionnaires will allow researchers more latitude in the comparison of actor and observer attributions.

Data Analysis

The SASQ is scored by calculating the mean scores on each dimension across all the scenarios. Each mean is a subscale of the questionnaire, representing one of the five attributional dimensions. Discriminant validity for each subscale was assessed by determining the intercorrelations between the subscales as well as through information provided by factor analyses.

RESULTS

The internal reliability of each subscale (or dimension) was evaluated by calculating Cronbach's Alphas for the sample of 304 usable responses. Three of the five dimensions had reliabilities that meet the .70 criterion established by Nunnally (1978): stability (alpha = .7781), globality (alpha = .7782), and intentionality (alpha = .8219). The LOC and controllability dimensions were not acceptably reliable (alphas = .5309 and .5137, respectively). Descriptive statistics, correlations, and reliabilities for each scale are reported in Table 9.2.

**Table 9.2.   Descriptive Statistics, Correlations,
and Scale Reliabilities for the SASQ: Study 1**

Dimension	Mean	sd	1	2	3	4	5
Locus of casuality	3.68	.68	(.5309)				
Stability	5.18	.65	–.0612 p = .150	(.7781)			
Globality	4.98	.80	–.0898 p = .064	.5637 p < .001	(.7782)		
Control	4.10	.66	–.5692 p < .001	.1584 p = .003	.2131 p < .001	(.5137)	
Intention	2.36	.77	–.1324 p = .012	–.1817 p = .001	–.1608 p = .003	.2257 p < .001	(.8219)

n=304

Factor Analyses

In order to assess the degree to which the SASQ measured the anticipated set of latent variables, a confirmatory factor analysis was conducted
using LISREL VIII software. The model tested specified five latent variables, each corresponding to one of the attributional dimensions. The
individual questions were specified to load on their corresponding latent
variable. The confirmatory analysis failed to find a satisfactory solution,
with only interim solutions being reported by the statistical software. Since
this indicates that the model is a poor fit, an exploratory analysis of the
data was performed. The initial exploratory analysis was done using an
oblique rotation (OBLIMIN) and extracted factors using the criteria that
the eigenvalue is greater than one, a commonly used criterion. The factor
analysis revealed several general factors that reflected the intent, stability,
and LOC dimensions, somewhat confounded by situational variables
(except for intent). LOC actually seemed to have a separate factor for most
individual scenarios, with a few representing a very small group of scenarios (two or three). There were a few multiple factors for stability, and intent
factored unidimensionally.

Given that the goal of the questionnaire is to measure the underlying
dimensions, and not the situational factors, a three-factor model was
forced, using a QUARTIMAX rotation, noted for its usefulness in identifying global or general factors (Kim & Mueller, 1978; Stephens, 1986). This
method was chosen for its ability to identify multiple moderate loadings on
each factor (Kim & Mueller, 1978), which is what would be expected in this
case. In addition, since the controllability variables had no independent

loadings, always loading with LOC, only the LOC variables were used, and for the same reason, only the stability variables were included from the stability/globality pair. Other reasons for dropping the controllability and globality dimensions include the fact that the scales were highly correlated with their LOC and stability counterparts. Also, in examining the correlations between the raw questionnaire items, it became apparent that the variables were highly correlated on a case-by-case basis as well. Furthermore, the controllability scale was not internally reliable. Finally, past research in attribution theory has suggested that these two dimensions may not exist independent of the LOC and stability scales. These reasons, with the factor analysis results, provide the rationale for dropping the controllability and globality dimensions from the analysis.

Two of the three factors extracted were easily interpreted, an intentionality factor and a stability factor. The third factor loaded on a few of the LOC questions, but did not appear to be a general factor. This could be related to the low internal reliability for this dimension. Therefore, one last analysis was run, specifying a two-factor solution with QUARTIMAX rotation. This analysis yielded two interpretable factors: stability and intentionality. These two factors had significant loadings (above .30) on all 16 of their corresponding variables. These findings are summarized in Table 9.3.

Table 9.3.   Rotated Two-Factor Solution for Study 1

	Factor 1 (Intent)	Factor 2 (Stability)
Stability 1	−.03837	**.31041**
Intentionality 1	**.46034**	−.00787
Stability 2	.00003	**.46652**
Intentionality 2	**.51451**	−.16487
Stability 3	−.04516	**.52478**
Intentionality 3	.56935	−.08918
Stability 4	−.14617	**.46928**
Intentionality 4	.34851	−.07872
Stability 5	.00110	**.49556**
Intentionality 5	**.48068**	.01939
Stability 6	−.12328	**.53080**
Intentionality 6	**.37508**	−.00447
Stability 7	.14076	**.49657**
Intentionality 7	**.47852**	−.08864
Stability 8	−.05132	**.57934**
Intentionality 8	**.66632**	−.03918

Table 9.3.   Rotated Two-Factor Solution for Study 1 (Cont.)

	Factor 1 (Intent)	Factor 2 (Stability)
Stability 9	−.04102	**.53778**
Intentionality 9	**.49889**	−.27973
Stability 10	.01392	**.43369**
Intentionality 10	**.41550**	−.04216
Stability 11	−.09013	**.50187**
Intentionality 11	.56478	−.07511
Stability 12	−.06562	**.48453**
Intentionality 12	**.43547**	−.04413
Stability 13	−.06307	**.41595**
Intentionality 13	.68290	.06396
Stability 14	.00845	**.53221**
Intentionality 14	.54931	−.10038
Stability 15	−.15869	**.52335**
Intentionality 15	**.58656**	.04159
Stability 16	−.00638	**.38909**
Intentionality 16	**.65673**	−.00876

Variables are listed by scenerio and dimension measured (e.g. stability 1 is question 2 on the instrument). **Significant loadings appear in bold print.**

While the total explained variance in the two-factor solution is lower than the earlier solutions (26.5%), this is the variance attributable to the underlying style of attributions in the absence of situational factors. This common variation can be said to be measuring the social attributional style that this study hypothesized to exist. These results suggest that observers do have attribution styles, consisting of at least two factors: stability and intentionality. Furthermore, the lower percentage of explained variance in the two-factor solution suggests that there is also a considerable situational component to observer attributions. We would like to stress that, since this is the first version of this questionnaire, it is possible that instrument problems contributed to the low reliability and lack of independence of the LoC and globality dimensions.

DISCUSSION

The presence of a social attribution style was confirmed in that two dimensions, stability and intentionality, were found to be present using factor analysis. Furthermore, reliabilities on these two scales were high, close to or above .8. This is particularly encouraging considering that the most widely used measure of self attributions, the ASQ, is reported to have reliabilities as low as .40 (Peterson et al., 1982).

These results suggest that, for future measurement of social attributions, the dimensions of stability and intentionality should be included. Furthermore, due to the somewhat mixed results on the locus of causality (LOC) dimension, it would seem prudent to include LOC as well. More specifically, although a single LOC factor was not found in the unconstrained exploratory analysis, LOC factors were extracted for a number of individual scenarios. Thus the locus of causality questions loaded highly and significantly on a single scenario-based factor for most scenarios. This suggests that while situational components may influence LOC attributions more than any cross-situational style, observers nevertheless make LOC distinctions. Thus our results do not indicate that the LOC dimension is not important; just that it appears to be influenced more heavily by the situation (explaining the separate factors) than by any cross-situational bias. Some of this may be due to the type of instrument used, or the character of the questions asked.

In conclusion, the current study demonstrates that there are cross-situational consistencies that influence individuals social attributions, at least on the questionnaire that we developed. Given the importance of evaluations in organizational contexts, coupled with the evidence that attributional biases affect leaders' evaluation of members (Dobbins & Russell, 1986; Martinko & Gardner, 1987), additional research appeared to be warranted and a second study was conducted in order to provide validation of the SASQ.

STUDY II

This study replicates and extends the prior study by further examining the factor structure of the SASQ and providing construct validation of the SASQ by examining the relationship between it and other established measures. As described above, in the first study we found strong support for two of the five potential dimensions: stability and intentionality. However, since the reliabilities for the other dimensions were low, there is the possibility that there may have been a sample specific measurement problem. In view of this possibility, we decided to also include the dimensions of inter-

nality, controllability, and globality in the SASQ for this study. The rationale for the theoretical relationships that we predicted for the SASQ are
described in the following paragraphs.

Self Attributions

As mentioned earlier, Martinko and Thomson (1998) contend that self
and social attributions are made using the same basic process. As a result, it
is possible and even probable that both SASs and self attribution styles may
be similar within individuals. Thus we propose:

Hypothesis 1:   *Individual dimensions on the SASQ will be positively related to
their OASQ counterparts.*

Two more hypotheses that may help in examining the construct validity
of the SASQ concern the actor-observer bias. This bias states that observers,
as contrasted with actors, tend to make attributions that are more internal
and stable for actor failures. One explanation for this bias is that the actors
focus on the situational determinants of their behavior whereas observers'
focus of attention is on the actor and the actor's characteristics, which are
viewed as internal and stable. Therefore, we would expect that, within the
same individuals:

Hypothesis 2:   *The SASQ scores will be significantly more internal than the OASQ
scores on the locus of causality dimension.*

Hypothesis 3:   *The SASQ scores for the stability dimension will be significantly
more stable than the scores for the OASQ.*

Theory X/Theory Y

McGregor (1960) suggested that managers have implicit theories about
the nature of people at work. He suggested two extremes, Theory X and
Theory Y. A person's general theory X/theory Y belief about workers is a
type of person schema, as person schemas are defined as "containing information about typical people or prototypes" (Brigham, 1991, p. 48). Since
schemas are used to predict behavior, managers who fall closer to the theory X end of the continuum would view failures as intentional, and managers who are closer to the theory Y end would view them as unintentional.
Furthermore, since theory X beliefs include low abilities of workers (dumb,
lazy, etc.), theory X managers would be more likely to make internal, stable
(ability) attributions for failures. Thus:

Hypothesis 4: *For failure events on the SASQ, theory X beliefs will be positively related to the intentionality dimension of the SASQ.*

Hypothesis 5: *For failure events on the SASQ, theory X beliefs will be positively related with the internality dimension of the SASQ.*

Hypothesis 6: *For failure events, theory X beliefs will be positively related to the stability dimension of the SASQ.*

Field Dependence

Field dependence refers to a person's ability to attend to a focal stimulus in the face of other compelling yet irrelevant stimuli (Witkin, Dyk, Faterson, Goodenough, & Karp, 1962). As Cardy and Kehoe (1984, p. 590) point out, field-independent people are more able to "separately attend to the dimensions of a multidimensional task." The field-dependent people, on the other hand, are "dominated by the configuration and have difficulty attending to the dimensions of a multi-dimensional stimulus." Allport (1960, p. 303) suggests that field-dependent people "jump to premature hypotheses and demand a definiteness in the outer world that it may not in fact possess." Therefore, they may be more inclined to use schemas, or some other sort of classification system, to relate events to this hypothesized order. Thus:

Hypothesis 7: *Field dependence will be related to the stability dimension of the social attribution style, such that field dependent individuals will generally make attributions of stable causes for failures.*

Leader Behaviors

A prominent model used in the study of leader behaviors is the Ohio State leadership studies model, which consists of two dimensions, initiating structure and consideration (Fleishman, 1957; Halpin & Winer, 1957; Hemphill & Coons, 1957). The initiating structure dimension refers to leader behaviors that define the relationship between leader and subordinate, and the degree to which a leader originates, facilitates, or resists new ideas and practices (Halpin & Winer, 1957; Hemphill & Coons, 1957). However, this dimension was later divided into three categories; role clarification, work assignment and specification of procedures (House & Dessler, 1974; Schriesheim, 1978).

Consideration, also called supportive leadership (House & Dessler, 1974), is characterized by behaviors that make a person "friendly and approachable, and considerate of the needs of subordinates" (p. 41). This variable has been found to be unconfounded and unidimensional (Schriesheim, 1978).

The behaviors identified in the previous section, initiating structure (all three types) and consideration, are expected to be related to the leader's attribution style. If a manager feels that an employee is lacking in ability or intelligence (internal, stable attribution), she would be likely to attempt to simplify the tasks that she assigns to that employee. She would tend to assign specific tasks and procedures, rather than giving more general goals. If she felt failure was due to lack of effort, or some other internal, unstable attribution, she may be likely to use these methods to a lesser degree. On the other hand, if she perceives that the cause of the failure was due to some random external occurrence, or the difficulty of the task, then she would not be likely to use the above-mentioned behaviors at all. Therefore, managers who generally attribute failures to internal and stable causes will be more likely to feel that directive behaviors (role clarification, specification of procedures, and work assignment) are necessary.

Consideration, on the other hand, may be related to intentionality, as unintentional failures may be more likely to result in supportive responses than a failure that is seen as intentional. The same is true for controllability, as an uncontrollable failure will evoke more supportive response than a controllable one. Thus we proposed:

Hypothesis 8:   *The role clarification, work assignment, and specification of procedures dimensions of leader behavior will be related to the locus of causality and stability dimensions of the SAS, such that high scores on these dimensions of leader behavior will positively be related to internal and stable attributions for failures.*

Hypothesis 9:   *The supportive leader behavior dimension of leader behavior will be related to the dimension of intentionality in the SAS such that high scores on supportive behavior will generally be related to attributions of unintentional causes for failures.*

Hypothesis 10:   *The supportive leader behavior dimension of leader behavior will be related to the dimension of controllability in the SAS such that high scores on supportive behavior will generally be related to attributions of uncontrollable causes for failures.*

METHODS: STUDY II

Sample

The sample consisted of 227 undergraduate students, enrolled at one of two universities in the southeastern United States. Of these, 204 returned usable responses, for a response rate of 89.4%. Demographic data for the sample are reported in Table 9.4. Descriptive statistics, correlations, and reliabilities are reported in Table 9.5. Cronbach's alphas for the four of the five scales of the SASQ were in the acceptable range of .70 suggested by Nunnally (1978). The controllability dimension was marginally reliable (alpha = .6176).

Table 9.4.   Sample Demographics for Study 2

Characteristic	Number	%
Sex		
Male	139	61.2
Female	88	38.8
Year		
Sophomore	1	.4
Junior	34	10.6
Senior	185	81.5
Graduate	7	3.1
GPA		
1.00–2.00	3	1.3
2.01–2.50	24	10.6
2.51–3.00	107	47.1
3.01–3.50	60	26.4
3.51–4.00	33	14.5
Work Experience		
None	8	3.5
1 Year	22	9.7
2 Years	15	6.6
3–5 Years	71	31.3
More than 5 Years	111	48.9
Managerial/Supervisory Experience		
Yes	113	49.8
No	106	46.7
Not Sure	7	3.1
Reported Academic Major		
Business	25	11.0

Table 9.4.   Sample Demographics for Study 2 (Cont.)

Characteristic	Number	%
Management	19	8.4
Human Resource Management	2	.9
Finance	43	18.9
Accounting	35	15.4
Marketing	36	15.9
Management Information Systems	22	9.7
All others (including dbl. majors)	45	19.8

Factor Analysis

The model developed in Study I, which specified stability and intentionality as the two latent variables, was tested with LISREL VIII software. The results again supported the two factor model. However, the chi-square indicated that the model was not an ideal fit (CHI-SQUARE with 463 degrees of freedom = 734.61 [P = 0.00]). However, the chi-square is sensitive to sample size and violations of normality (Stevens, 1986). Since there is skew to some of the variables, particularly the Intentionality variable (see Table 9.2), and the sample size is small relative to the number of variables, this concern is relevant. As Joreskog and Sorbom (1993) explain,

> The use of the chi-square as a central χ^2 statistic is based on the assumption that the model holds exactly in the population. As already pointed out, this may be an unreasonable assumption in most empirical research. A consequence of this assumption is that models which hold approximately in the population will be rejected in large samples. (p. 123)

Browne and Cudeck (1993) suggested that a better statistic is Steiger's (1990) root mean squared error of approximation (RMSEA). They further suggested that a value on the RMSEA of ".05 or lower indicates close fit and values up to .08 represent reasonable errors of approximation in the population" (Joreskog & Sorbom, 1993, p. 124). For this two-factor model the RMSEA = 0.054. This indicates that while the fit is not excellent, it is reasonable. (See Table 9.6 for factor loadings.)

However, since the reliability of the locus of causality dimension was higher in this study than in Study I, a three-factor model was also run. The chi-square was again significant, and once again the RMSEA indicated adequate fit (RMSEA = 0.059), however the goodness of fit index (GFI) decreased from .82 to .74 . This indicates that the data better represent a two-factor model consisting of stability and intentionality. Further support

Table 9.5. Factor Loadings Using LISREL

	Stability	Intentionality
Stability 1	0.49	—
Intentionality 1	—	0.29
Stability 2	0.44	—
Intentionality 2	—	0.40
Stability 3	0.52	—
Intentionality 3	—	0.56
Stability 4	0.44	—
Intentionality 4	—	0.51
Stability 5	0.37	—
Intentionality 5	—	0.51
Stability 6	0.40	—
Intentionality 6	—	0.40
Stability 7	0.40	—
Intentionality 7	—	0.43
Stability 8	0.49	—
Intentionality 8	—	0.64
Stability 9	0.52	—
Intentionality 9	—	0.67
Stability 10	0.50	—
Intentionality 10	—	0.56
Stability 11	0.51	—
Intentionality 11	—	0.56
Stability 12	0.47	—
Intentionality 12	—	0.63
Stability 13	0.53	—
Intentionality 13	—	0.59
Stability 14	0.55	—
Intentionality 14	—	0.63
Stability 15	0.51	—
Intentionality 15	—	0.47
Stability 16	0.36	—
Intentionality 16	—	0.56

Root mean square of approximation (RMSEA) = .0540
n = 203

Table 9.6.   Descriptive Statistics, Correlation and Scale Reliabilities for the SASQ

Dimension	Mean	sd	1	2	3	4	5
Locus of Causality	5.23	.82	(.7170)				
Stability	5.46	.72	.3245	(.8145) p<.001			
Globality	5.32	.77	.2815	.6326 p < .001	(.7649) p < .001		
Controllability	4.32	.79	.3285	.0698 p < .001	.1299 p = .161	(.6176) p = .032	
Intent	2.55	.95	−.0866	−.2678 p = .012	−.1910 p < .001	.1911 p = .003	(.8575) p = .003

n=203

for the construct will require the linking of these dimensions to other existing constructs. This will be done in the following section.

Hypothesis Testing

Due to the number of hypotheses being tested, and the techniques used to test them, the hypotheses will be examined individually, or in small groups. H1, which proposed that the SASQ would be positively related to its OASQ counterparts, was supported (see Table 9.7). There were significant correlations between the LOC dimensions (.6886, $p < .001$) and the

Table 9.7.   Correlations between the SASQ and OASQ

	OASQ Dimensions	
SASQ Dimensions	Locus of Causality	Stability
Locus of Causality	.6886***	.0457
Stability	.1866	.5630***
Globality	.1061	.3454***
Controllability	.2311**	−.1486*
Intentionality	−.0226	−.0224

Significant at alpha = .05 *
.01 **
.001 ***

stability dimensions (.5630, $p < .001$). There were also significant correlations between the OASQ stability dimension and SASQ globality dimension (.3454, $p < .001$) and between OASQ locus of causality and SASQ controllability dimensions (.2311, $p < .05$).

To test H2 and H3, which predicted that the SASQ scores would be more internal and stable than the OASQ scores, t-tests for related samples were conducted on the paired difference scores testing the null hypothesis that the means are equal. For the first pair the SASQ LOC (mean = 5.08) and the OASQ LOC (mean = 5.23), the difference was not significant at the .05 level ($t = 1.31$, $p = .0951$), although it was in the predicted direction. For the stability scores, the difference was significant and in the predicted direction ($t = 6.53$, $p < .001$).

The remainder of the hypotheses involve other instruments hypothesized to be related to SASQ. A multivariate analysis of variance (MANOVA) was first conducted using all of the variables, and the multivariate test statistic was used as a control against inflation of error rates. This statistic (Wilk's lambda = .19913, n = 155, df = 10,144) was significant (p is less than .0001), indicating that further testing was warranted.

The remaining hypotheses were sequentially examined using MANOVAs on each set of related data. This was done in this manner for a number of reasons. First, since some of the variables were independent variables affecting SASs, some were dependent variables affected by SAS, and some are covariates, analyzing these variables simultaneously would present difficulty in interpretation of results. Furthermore, a loss of power would result from doing a single MANOVA, as not all subjects completed the whole questionnaire.

The MANOVA's all had sample sizes over 180; however, the overall test had a sample size of 155. This is because different subjects omitted different sections of the questionnaires. Finally, this does allow some control for inflated error rates within each MANOVA. The summary statistics for the other instruments used are reported in Table 9.8.

Table 9.8. Descriptive Statistics for Scale Scores on Comparison Instruments

Scale	Mean	Std. Dev.	Min	Max
OASQ				
Locus of Causality	5.08	1.27	1.25	7.00
Stability	4.83	1.05	2.19	7.00
Theory X/Y Questionnaire	2.56	.41	1.50	3.71
Group Embedded Figures Test (Field Dependence)	11.46	4.98	0.00	18.00

**Table 9.8. Descriptive Statistics for Scale
Scores on Comparison Instruments (Cont.)**

Scale	Mean	Std. Dev.	Min	Max
Leader Behavior Questionnaire				
Work Assignment	3.97	.52	2.13	5.00
Specification of Procedures	3.48	.67	1.00	5.00
Role Clarification	4.59	.48	2.75	5.00
Supportive Behavior	4.12	.56	1.38	5.00

Because of the early stages of development of this instrument, and the theory itself, tested in our study, we included all five dimensions of the SASQ in our analyses of the remaining hypotheses, in spite of the low reliabilities of some of the dimensions. As stated earlier, this is done since we don't know for certain whether these results indicate the actual lack of a LOC and globality dimension, or merely weaknesses in the testing instrument.

H4, 5, and 6, which predicted positive relationships between Theory X beliefs and the intentionality, internality, and stability dimensions of the SASQ, were examined simultaneously using MANOVA. The Wilk's lambda was .94964 n = 199, (1,197)df. This was not significant at the .05 level (p = .074). However, since this was a borderline case (p is less than .10), the univariate tests were examined, and there was a significant relationship between Theory X beliefs and intentionality (p = .004), suggesting that people with Theory X beliefs are more likely to attribute failures to intentional causes.

A MANOVA was conducted with all five SASQ dimensions as dependent variables, and Field Dependence scores as independent variables to test H7. The multivariate test was significant (Wilk's lambda = .93926, n = 183, (1,181)df, p = .048). However, the hypothesized relationship between stability and field dependence was not significant in univariate testing (f = 2.03823, p = .155). There was, however, a significant negative relationship between field dependence scores and attributions of intentionality (f = 7.121, p = .008). High scorers on the field dependence instrument (field-independent people) attribute failures to less intentional causes than field-dependent people.

Hypotheses 8–10, which all relate leader behaviors to SASQ scores, were tested simultaneously using MANOVA. The Wilk's lambda (.76992) was significant (p < .001). Univariate tests indicated positive relationships between role clarification and the locus of causality (B = .40685, p = .006), between role clarification and stability and controllability (p's = .042, .036., respectively), and between work assignment and globality (p = .020). A negative relationship was also present between role clarification and the intent dimension. Thus, there was general support for the proposition that SASs are related to leader behaviors.

DISCUSSION

Overall, the results of both Study I and Study II support the idea of coherent and reliable SASs, suggesting that people have consistent preferences for the types of attributions that they make when they observe the outcomes of others. More specifically, based on the current findings, the evidence points to the existence of a cross-situational style of social attributions, having at least two dimensions, stability and intentionality. Furthermore, there is a strong situational impact on locus of causality attributions and we expect that in some situations, particularly where the observer knows the actor, the locus of causality dimension may appear as a significant factor. The remaining two dimensions of attributions, globality and controllability, were not supported as independent dimensions, as they loaded on the locus of causality and stability (respectively) dimensions during factor analysis. Thus globality and controllability should probably not be included in future scales since they are too highly correlated with stability and locus of causality (respectively) to be considered separate dimensions.

There was also reasonable support for the construct validity of the SASQ. In general, there was evidence that the SASQ, and SASs in general, are related to self-attribution styles, Theory X beliefs, field dependence, and leader behaviors. More specifically, as we discussed above, since we postulated that self and social attributions are a reflection of the same fundamental process, we expected that the OASQ and SASQ scales developed to measure attribution styles would be correlated to some degree, and they were. The finding that they were not perfectly correlated adds to the discriminant validity of the SASQ.

In addition, based on our general knowledge of the self-serving bias, we expected that the attributions for the failures of others elicited by the locus of causality and stability dimensions of the SASQ would be more internal and stable than the attributions for self failures elicited by the OASQ. We found that the differences were in the predicted direction and significant with respect to the stability dimension. Thus, the findings regarding the types of similarities and differences we found between the two scales appear to be consistent with existing theory, adding to the construct validity of both scales.

We also verified several of the other relationships that were predicted, including the relationship between Theory X beliefs and intentionality, and the relationships between the leader behaviors of role clarification and locus of causality, stability, and controllability.

Other relationships that were found, although not predicted, appear to be consistent with theory. For example, the finding that high scorers on the field dependence instrument (field-independent people) attribute failures to less intentional causes than field-dependent people appears to make sense in that, based on the characteristics of people who are field independent, we would expect that their focus of attention would be more eclectic and include the actor as well as the actor's environment. On the other hand, it appears more likely that field-dependent people would focus most of their attention on the actor. As a result of these differences in perceptual focus, it makes sense that field-independent people made less intentional attributions for actor behavior, since their focus of attention also includes the environmental context while the focus of attention of field-dependent people does not include the environment. Similarly, based on our understanding of leadership and attribution theory, one would expect that the people who scored high on role clarification behaviors would not attribute failure to intentional causes, since they would believe that such behaviors could be corrected through leader behavior associated with role clarification.

In summary, it appears that SAS, as characterized by the SASQ, is a reliable construct with at least two dimensions and that it is a valid construct in that it has significant relations with other constructs that are anticipated by theory. Thus, considering the exploratory nature of this study, these results are promising.

It is appropriate to mention several limitations and reservations regarding the two studies and their results. First, the sample was comprised of student subjects and, although most had some work experience, they were undoubtedly less experienced and younger than the average member of the workforce. Although we would not anticipate fundamental differences in the construct of attributions style with an older and more experienced sample, constructive replications would increase the confidence in our findings. Second, the studies involved the relationships between multiple questionnaires, most of which were Likert-type, and therefore the potential for common-method variance cannot be eliminated. We stress that this research was meant to probe into a new area, and as such will require further studies, perhaps using a different method, to further validate the theory.

Another aspect of our study affecting the validity of the construct of SASs is that the SASQ focuses on work-related situations. As discussed earlier, this was viewed as an asset since research has suggested that styles are more cohesive within definable behavioral domains and that scales that are more broad in scope generally have lower reliabilities (Kent & Martinko, 1995). However, this restriction of the domain also has some liabilities in that, although our research supports the construct of a work-related SAS, a more general construct of SAS has not yet been validated. Again, we anticipate that SASs affect a larger domain than work-related attributions, but additional research in other behavioral domains would certainly strengthen the general construct of SAS.

Some comments regarding the application of SAS and future research strategies are warranted. A major implication of SAS is that such styles would affect leader observations and evaluations of members, causing leaders to attribute either more or less responsibility for failures to members than is warranted. Thus, SASs are expected to play a large role in performance appraisal situations and should also be related to both perceptions and behaviors associated with organizational justice (Greenberg, 1990). Consequently, research designed to assess the effects of SASs in the areas of performance appraisal and organizational justice would appear to be particularly beneficial by providing another level of explanation for the perceptual errors that are associated with these processes and by contributing to the construct validity of SAS.

In conclusion, our data and results clearly demonstrate that SASs exist in work-related domains and can be characterized by the dimensions of stability and intentionality, and perhaps by locus of causality. The data also contribute to the construct validation of SASs, indicating that SASs affect or are affected by other constructs including self attribution styles, self-serving biases, Theory X, field dependence, and leader behavior. However, it should be noted that the dimensions that were not supported by these studies, locus of causality and globality, may have been present, but merely not measured accurately by the instrument. Therefore, one of the future directions for this research is to further develop the measurement instrument in order to determine whether or not these dimensions exist in a SAS. Note that future lack of support for these dimensions would not mean that those dimensions are irrelevant for social attributions, but simply that they are so strongly situationally bound that they don't reflect a cross-situational stability or style. We believe that the construct of SAS provides a potentially promising explanation for some of the perceptual problems associated with performance appraisal processes and perceptions of organizational justice. Additional research assessing the generality of the construct of SAS is warranted. Additional construct validation would be helpful, particularly in the areas of appraisal and organizational justice.

APPENDIX

Social Attribution Style Questionnaire

Instructions: In this questionnaire, a number of situations are presented. Read each situation and imagine it happening to a coworker of yours in your current job or the last job that you held in which the situation is applicable. Based on what you know about people in general, write down what you think most likely to be the one major cause of the event in the space provided. Respond to each of the questions that follow the event by circling the number on the scale that best describes the cause you identified.

Items

1. Your Coworker recently received a below-average performance evaluation from their supervisor.
2. Today you are informed that suggestions made by a worker in your organization to your boss in a recent meeting would not be implemented.
3. A coworker complains that they will not receive a promotion that they have wanted for a long time.
4. You recently discovered that a coworker of yours is being paid considerably less than another employee holding a similar position.
5. You find out that another employee failed to achieve all of their goals for the last period
6. A fellow employee has a great deal of difficulty getting along with their coworkers.
7. A customer recently complained about the service another employee at your firm provided them.
8. A coworker was not selected for advanced training that they had expressed a strong desire to attend.
9. A large layoff has been announced at your company, and a worker in your unit was told that they will be one of those laid off.
10. You just learned that a coworker will not be reimbursed for expenses they recently submitted.
11. A new member of your work group is having a great deal of difficulty learning how to use the new computer.
12. You find that another employee recently received a below-average raise.

13. All of the feedback your boss has given another worker lately concerning their performance has been negative.

14. A coworker was not nominated by their peers for a special award that they would like to receive.

15. Your boss does not take one of your peers seriously.

16. There is a serious accident at work involving one of your coworkers.

Scales

For each item:
Write down the one major cause _____

1. To what extent is this cause due to something about this coworker or something about other people or circumstances ?

   completely   completely due to
   due to   1 2 3 4 5 6 7   other people
   coworker   or circumstances

2. Will this cause be present in future situations that are similar?

   never   always
   present   1 2 3 4 5 6 7   present

3. Is this cause something that affects just this type of situation, or does it affect other situations at work?

   Just this type of situation   All types of
   of situation 1 2 3 4 5 6 7   situations

4. To what extent is this cause something that this coworker has under their control?

   Not at all   Completely under
   Under their 1 2 3 4 5 6 7   their control
   control

5. To what extent is this cause something that they intended to have happen?

   Not what   Exactly what
   they intended 1 2 3 4 5 6 7 they intended

REFERENCES

Abramson, L. Y., Seligman, M. E. P., & Teasdale, J. D. (1978). Learned helplessness in humans: critique and reformulation. *Journal of AbnormalPsychology, 87*, 49–74.

Adler, S. (1980). Self-esteem and causal attributions for job satisfaction and dissatisfaction. *Journal of Applied Psychology, 65*, 327–332.

Allport, G. W. (1960). *Personality and social encounters.* Boston: Beacon Press.

Anderson, C. (1983). The causal structure of situations: The generation of plausible causal attributions as a function of type of event situation. *Journal of Experimental Social Psychology, 19*, 185–203.

Anderson, L. R. (1992). Leader interventions for distressed group members: Overcoming leaders' self-serving attributional biases. *Small Group Research, 23*(4), 503–523.

Ashkanasy, N. (1989). Causal attributions and supervisors' response to subordinate performance: The Green and Mitchell model revisited. *Journal ofApplied Social Psychology, 19*, 309–330.

Ashkanasy, N. (1995). Supervisory attributions and evaluative judgments of subordinate performance: A further test of the Green and Mitchell model. In M. J. Martinko, (Ed.), *Attribution theory: An organizational perspective* (pp. 211–228). Delray Beach, FL.: St. Lucie Press.

Bettman, J., & Weitz, B. (1983). Attributions in the board room: Causal reasoning in corporate annual reports. *Administrative Science Quarterly, 28*, 165–183.

Blakely, G. L. (1993). The effects of performance rating discrepancies on supervisors and subordinates. *Organizational Behavior and Human DecisionProcesses, 54*, 57–80.

Brigham, J. C. (1991). *Social psychology* (2nd ed.). New York: HarperCollins.

Campell, C. C., & Martinko, M. J. (1990) An integrative attributional perspective of empowerment and learned helplessness: a multimethod field study. *Journal of Management, 24*(2), 173–200.

Cardy, R. L., & Kehoe, J. F. (1984). Rater selective attention ability and appraisal effectiveness: the effect of a cognitive style on the accuracy of differentiation among ratees. *Journal of Applied Psychology, 69*, 589–594.

Christie, R. (1970). Relationships between machiavellianism and measures of ability, opinion and personality. In Christie & Geis (Eds.), *Studies in Machiavellianism* (pp. 35–52). New York: Academic Press.

Clapham, S. E., & Schwenk, C. R. (1991). Self-serving attributions, managerial cognition, and company performance. *Strategic Management Journal, 12*(3), 219–229.

Cronshaw, S. F., & Lord, R. G. (1987). Effects of categorization, attribution, and encoding processes on leadership perceptions. *Journal of Applied Psychology, 1*, 97–106.

Cutrona, C. E., Russell, D., & Jones, R. D. (1985). Cross-situational consistency in causal attributions: does attributional style exist? *Journal of Personality and Social Psychology, 47*, 1043–1058.

Deiner, C. T., & Dweck, C. S. (1978). An analysis of learned helplessness: Continuous changes in performance, strategy, and achievement cognitions following failure. *Journal of Personality and Social Psychology, 36*, 451–462.

Dobbins, G. H., & Russell, J. M. (1986). The biasing effects of subordinate likeableness on leaders' responses to poor performance: A laboratory and a field study. *Personnel Psychology, 39,* 759–777.

Dykman, B., & Abramson, L. (1990). Contributions of basic research to the cognitive theories of depression. *Personality and Social Psychology Bulletin, 16*(1), 42–57.

Fedor, D. B., & Rowland, K. M. (1989). Investigating supervisor attributions of subordinate performance. *Journal of Management, 15*(3), 405–416.

Ferris, G. R., Yates, V. L., Gilmore, D. C., & Rowland, K. M. (1985). The influence of subordinate age on performance ratings and causal attributions. *Personnel Psychology, 38,* 545–557.

Fleishman, E. A. (1957). A leader behavior description for industry. In R. Stogdill & A Coons (Eds.), *Leader behavior: Its description and measurement.* Columbus, OH: Bureau of Business Research.

Green, S. G., & Liden R. C. (1980) Contextual and Attributional Influences on Control Decisions. *Journal of Applied Psychology, 65,* 453–458.

Greenberg, J. (1990). Organizational justice: yesterday, today, and tommorrow. *Journal of Management, 16,* 399–432.

Halpin, A. W., & Winer, B. J. (1957). A factorial study of the leader behavior descriptions. In R. Stogdill & A. Coons (Eds.), *Leader behavior: Its description and measurement.* Columbus, OH: Bureau of Business Research.

Hemphill, J. K., & Coons, A. E. (1957). Development of the leader behavior description questionnaire. In R. Stogdill & A. Coons (Eds.), *Leader behavior: Its description and measurement.* Columbus, OH: Bureau of Business Research.

Heneman, R. L., Greenberger, D. B., & Anonyuo, C. (1989) Attributions and exchanges: The effects of interpersonal factors on the diagnosis of employee performance. *Academy of Management Journal, 32*(2), 466–476.

Hesketh, B. (1984). Attribution theory and unemployment: Kelley's covariation model, self-esteem and locus of control. *Journal of Vocational Behavior, 24,* 94–109.

House, R. J., & Dessler, G. (1974). The path-goal theory of leadership: some post hoc and a priori tests. In J. G. Hunt & L. L. Anderson (Eds.), *Contingency approaches to leadership* (pp. 29–55). Carbondale: Southern Illinois University Press.

Joreskog, K., & Sorbom, D. (1993). *LISREL 8: Structural Equation Modeling with The Simplis Command Language.* Hillsdale, NJ: Erlbaum.

Kelley, H. H. (1973). The process of causal attribution. *American Psychologist, 28,* 107–128.

Kent, R. L., & Martinko, M. J. (1995). The development and evaluation of a scale to measure organizational attribution style. In M. J. Martinko (Ed.), *Advances in attribution theory: An organizational perspective* (pp. 53–75). Delray Beach, FL: St. Lucie Press.

Kim, J.-O., & Mueller, C. W. (1978). *Factor analysis: Statistical methods and practical issues.* Newbury Park, CA: Sage.

Klaas, B.S. and Wheeler, H.N. (1990). Managerial decision making about employee discipline: A policy-capturing approach. *Personnel Psychology, 43,* 117–134.

Knowlton, W. A., & Mitchell, T. R. (1980) Effects of causal attributions on a supervisor's evaluation of subordinate performance. *Journal of Applied Psychology, 65,* 459–466.

Liden, R. C., & Mitchell, T. R. (1985) Reactions to feedback: The role of attributions. *Academy of Management Journal, 28*(2), 291–308.

Martinko, M. J. (1995). The nature and function of attribution theory within the organizational sciences. In M. J. Martinko (Ed.), *Attribution theory: An organizational perspective* (pp. 7–16). Deray Beach, FL: St. Lucie Press.

Martinko, M. J., & Gardner, W. L. (1987). The leader/member attribution process. *Academy of Management Review, 12*(2), 235–249.

Martinko, M. J., & Thomson, N. (1998). A synthesis of the Kelley and Weiner attribution model. *Basic and Applied Social Psychology, 20*(4), 271–284.

McGgegor, D. (1960). *The human side of enterprise.* New York: McGraw-Hill.

Mitchell, James V., Jr. (1989) Personality correlates of attributional style. *Journal of Psychology, 123*, 447–463.

Mitchell, T. R., & Kalb, L. S. (1982). Effects of job experience on supervisor attributions for a subordinate's poor performance. *Journal of Applied Psychology, 67*, 181–188.

Mitchell, T. R., & Wood, R. E. (1980) Supervisor's responses to subordinate poor performance: A test of an attributional model. *Organizational Behavior and Human Performance, 23*, 429–458.

Mone, M. A., & Baker, D. D. (1992). Cognitive, affective, and behavioral determinants and consequences of self-set goals: An integrative, dynamic model. *Human Performance, 5*(3), 213–234.

Moss, S., & Martinko, M. (1997). The effects of performance attributions and outcome dependence on leader feedback behavior following poor performance. *Journal of Organizational Behavior, 19*, 259–274.

Nunnally, J. (1978). *Psychometric theory.* New York: McGraw-Hill.

Peterson, C., Semmel, A., Von Baeyer, C., Abramson, L., Metalsky, G., & Seligman E. (1982). The Attributional Style Questionnaire *Cognitive Therapy and Research, 6*, 287–300.

Peterson, C., & Villanova, P. (1988). An expanded Attributional Style Questionnaire. *Journal of Abnormal Psychology, 97*(1), 87–89.

Russell, D. (1982). The Causal Dimension Scale: A measure of how individuals perceive causes. *Journal of Personality and Social Psychology, 42*(6), 1137–1145.

Salancik, G. R., & Meindl, J. R. (1984). Corporate attributions as strategic illusions of management control. *Administrative Science Quarterly, 29*(2), 238–254.

Schriesheim, C. A. (1978). *Development, validation and application of new leadership behavior and expectancy research instruments.* Doctoral dissertation, Ohio State University.

Seligman, M., & Schulman, P. (1986). Explanatory style as a predictor of productivity and quitting among life insurance agents. *Journal of Personality and Social Psychology, 50*(4), 832–838.

Seligman M. E. P. (1990). *Learned optimism.* New York: Knopf.

Snyder, M. (1974). Self-monitoring of expressive behavior. *Journal of Personality and Social Psychology, 30*(4), 526–537.

Stevens, J. (1986). *Applied multivariate statistics for the social sciences.* Hillsdale, NJ: Erlbaum.

Thomson, N., & Martinko, M. (1995). The relationship between MBTI types and attributional style. *Journal of Psychological Type, 35*, 22–30.

Weiner, B. (1985). An attributional theory of achievement motivation and emotion. *Psychological Review, 92*(4), 548–573.

Witkin , H. A., Dyk R. B., Faterson, H. F., Goodenough, D. R., &Karp, S. A. (1962). *Psychological differentiation.* New York: Wiley.

Wrightsman, L. S., Jr., & Cook S. W. (1965). Factor analysis and attitude change. *Peabody Papers in Human Development, 3*(2).

CHAPTER 10

FOLLOWER ATTRIBUTIONS OF LEADER MANIPULATIVE AND SINCERE INTENTIONALITY

A Laboratory Test of Determinants and Affective Covariates

Marie T. Dasborough
University of Queensland

Neal M. Ashkanasy
University of Queensland

ABSTRACT

We examined in a laboratory study determinants and covariates of followers' attributions of their leader's intentions. We anticipated that follower perceptions of leader behavior, frame of reference (target or bystander), work experience, and mood would determine followers' attributions of their leader's manipulative and sincere intentions, and that followers' emotional reactions would covary with attributions. One hundred thirty-seven participants viewed a video of a transformational leader, and then received a simulated e-mail indicating the leader's self or organization focus, and the follower's frame of reference (target of influence or bystander). With the exception of frame of

Attribution Theory in the Organizational Sciences, pages 203–224

reference, results supported our predictions, further supporting the attributional theory of leadership.

The attributional theory of leadership has been researched extensively since its introduction in 1979 by Green and Mitchell, and has proved a valid and useful model of the leadership process (see, e.g., Ashkanasy, 1995; Ashkanasy & Gallois, 1994; Mitchell, Green, & Wood, 1981). More recently, however, Dasborough and Ashkanasy (2002) extended the model to incorporate notions of follower attributions of intentionality. Based on a more general model of intention attribution proposed by Ferris, Bhawuk, Fedor, and Judge (1995), Dasborough and Ashkanasy posited that followers' affective and behavioral reactions to perceived leader behaviors are determined in part by attributions as to whether the leader is motivated by manipulative motives that are likely to be detrimental to the organization, or sincere motives that are more likely to benefit the organization.

In the study we report here, we describe the first empirical tests of the central propositions in the Dasborough and Ashkanasy (2002) model. In this study, which was conducted in a laboratory setting, we focused on the antecedents and covariates of follower attributions of leaders' manipulative versus sincere intentions. Our model and hypotheses are based on a theory that links attributions of intentionality to perceptions of transformational leadership attempts and emotional reactions. We detail the theory behind these notions in the following paragraphs.

ATTRIBUTIONS OF LEADER INTENTIONALITY

According to Ferris and his colleagues (1995), the causes of behavior (i.e., intentions or motives) are central to attribution theory. In this respect, attribution theory is predicated on the idea that people search for the causes of their own and others' behaviors (Heider, 1958). Ferris and colleagues posit further that how we perceive others' motives for behaviors has significant impact on follower reactions to those behaviors.

Based on the Ferris and colleagues (1995) model, Dasborough and Ashkanasy (2002) developed a new attributional model of leadership, where follower attributions about leader motivations and intentions provide the means by which they differentiate between "true transformational" (sincere) and "pseudo-transformational" (manipulative) leaders. In this model, the perceived motives and intentions of the leader determine follower reactions to leadership behaviors, rather than any set of "objective" leader characteristics. In this sense, attributions regarding motives or intentions represent a triggering mechanism for differentiation of follower responses to the leader's influence attempts.

TRUE AND PSEUDO-TRANSFORMATIONAL LEADERSHIP

According to Owen (1986), true transformational leaders influence their followers positively through the management and skilful utilization of human resources. As Ashkanasy and Tse (2000) note, the aim of these leaders is to transform their organization through enticing followers to join them in achieving their visionary goals; and motivating them to behave in a way that contributes to their overall organizational plan. Howell (1988) has commented further that true transformational leadership is perceived positively by followers because it involves an external orientation; followers perceive the leaders' intention as serving the organization rather than serving themselves.

Weierter (1997) has noted, on the other hand, that transformational leaders may be destructive if they are self-serving, manifesting an internal focus. This negative side of transformational leadership has been characterized by Conger (1990) as "the dark side of leadership." Bass, Avolio, and Atwater (1996) coined the term "pseudo-transformational leadership" to describe manipulative, self-serving leadership. Pseudo-transformational leaders are thus perceived to use their transformational skills to manipulate followers in a controlling manner considered to be insidious or unfair by the follower (see also Owen, 1986). These leaders use their transformational skills (e.g., impression management, ability to formulate and communicate a vision) to exaggerate their claims for a vision, or to manipulate the audience viewing the message. According to Conger, the key indicator of pseudo-transformational leadership behavior is that followers perceive their leaders to be acting essentially for personal gain, rather than acting in the interests of the organization as a whole.

TRANSFORMATIONAL LEADERSHIP AND EMOTION

Humphrey (2002) has posited that leadership is intrinsically an emotional process. This is because leaders recognize their employees' emotional states; they attempt to evoke employee emotions and they seek to manage employees' emotions appropriately (see also Ashforth & Humphrey, 1995; Ashkanasy & Tse, 2000).

Transformational leadership, in particular, has been associated with emotions. Transformational leadership is largely explained by charisma (Ashkanasy & Tse, 2000; Bass, 1998), which enables leaders to control their own and others' emotions (see also George, 2000). This can have a unique impact on followers, such that they feel strongly affiliated with the leader and have an unquestionable willingness to obey the leader's instructions (Conger, 1990).

In a qualitative study of leader and follower perceptions of leadership, Dasborough and Ashkanasy (2003) provided additional insights into the emotional process of leadership. Followers were asked about positive and negative emotions that they had experienced in response to leader behaviors. Findings indicated that, when leaders displayed behaviors their followers believed were inappropriate or unexpected, the followers experienced negative emotions such as anger, frustration, and irritability. These conclusions, however, were based on data from recollections of past incidents, and only examined followers who were targets of leader behaviors. Furthermore, while the results of this study were intriguing, they provided little explanation of the antecedents of attribution formation or the cognitive processes that underlie follower reactions to transformational leadership influence attempts.

ATTRIBUTION AND EMOTION

Weiner (1977) was the first to note that causal attributions and emotional outcomes are linked. Weiner argued that locus of causality influences the affective or emotional consequences of achievement outcomes. Emotions are thus considered reactions to interpersonal evaluations (Weiner, Graham, & Chandler, 1982). For example, if an actor makes an internal attribution about a failure, such as "I failed because I did not put in enough effort," this is likely to be associated with feelings of guilt. In the case of success, if one makes an external attribution, such as "I succeeded because someone helped me", the emotional association is likely to be with gratitude.

Smith, Haynes, Lazarus, and Pope (1993) note that attributions of intentionality are more directly relevant to emotion than locus of causality. Attributions of intentionality are related to emotions such as guilt and shame (if we control the outcomes); or anger, gratitude, and pity if someone else controlled the outcome. Thus, attributions of intentionality are closely related to attributions of controllability. There is, however, a key distinction between the two. Weiner (1985) explains "the differentiation between intent and control lies at the heart of the distinction between murder and manslaughter" (p. 554). Examples of studies on attributions of intentionality and emotional outcomes are by Leon and Hernandez (1998) and Betancourt and Blair (1992). Nonetheless, no one to date has examined outcome-dependent affect, as seen in Weiner's (1985) attributional theory of motivation and emotion, prior to the formation of attributions of intentionality.

One aim of the present study therefore was to explore the effect of mood on perceptions of leadership, and the relationship between emotional reactions and attributions of intentionality following leadership influence attempts. The hypotheses to be tested in this study are based on propositions outlined by Dasborough and Ashkanasy (2002) in their model of follower attributions of leader intentionality, with a central focus on the role of emotion and the antecedents of follower attributions of intentionality. We investigated four groups of antecedents to attributions: leader behavior, follower frame of reference, follower experience, follower mood; and one covariate: follower emotional reactions to the leader's behavior.

Leader Behavior

In the case of highly charismatic transformational leaders, followers may be skeptical of their intentions due to their unique leadership style. "Too much of a good thing" (Baron, 1989) or "laying it on too thick" (Jones, 1990) often results in leader intentions being perceived by followers as questionable and self-enhancing. According to Ferris and colleagues (1995), situational norms can result in some influence attempts being perceived as genuine as opposed to being perceived as deceitful or manipulative. The display of transformational leadership behaviors will affect follower attributions of intentionality differently depending on situational norms. Given that transformational leaders are usually associated with organization-focused behaviors (Bass, 1998), if a charismatic transformational leader does not act in line with this norm, followers will be extra suspicious of the leader's behavior. Thus,

Hypothesis 1: *(a) Followers observing organizationally focused leadership behaviors will be more likely to attribute the behavior to sincere intentions. (b) Followers observing self-focused behaviors will be more likely to attribute the behavior to manipulative intentions.*

FOLLOWER FRAME OF REFERENCE

A perceiver, in this case the follower, may be a target or bystander of the leader's behavior, depending on the degree to which they are personally affected by the actor's behavior (Ferris et al., 1995). Jones (1990) explains that bystanders tend to be removed from the behavior and are therefore more critical of the actor's intentions. Thus, if perceivers are bystanders, they are more likely to make attributions of self-interested, manipulative intentions by the actor (Ferris et al., 1995). Following from this,

Hypothesis 2: *(a) Followers who perceive they are a target of the leader's transformational behavior are more likely to attribute leader behavior to sincere intentions. (b)*

Followers who see themselves as bystanders to the leader's transformational behavior will be more likely to attribute leader behavior to manipulative intentions.

FOLLOWER WORK EXPERIENCE

Ferris and colleagues (1995) suggest that a more experienced perceiver will be more likely to make attributions of deceitful intentions than a less experienced one, because those with less experience are more naive and susceptible to image management tactics. Organizational work experience will therefore influence how followers perceive their leaders' behavioral intentions. Thus,

Hypothesis 3:   *(a) Followers with less work experience will be less likely to make attributions of manipulative intentions than more experienced followers, (b) Followers with less work experience will be more likely to make attributions of sincere intentions than more experienced followers.*

FOLLOWER MOOD

Dasborough and Ashkanasy (2002) suggest that the positive or negative mood of the follower observing the leader behavior will influence how the follower initially perceives that behavior. Mood is defined as low intensity, relatively enduring affective states, without a salient antecedent cause and therefore very little cognitive content (Forgas, 1992). Mood is thus distinguished from emotion by the lack of a specific object causing the affective state. Forgas and George (2001) suggest that moods provide the underlying affective context for most of our behaviors and ongoing thought processes. Therefore, we hypothesize:

Hypothesis 4:   *(a) When followers are in a positive mood, they are more likely to attribute transformational leader behavior to sincere intentions. (b) When followers are in a negative mood, they are more likely to attribute transformational leader behavior to manipulative intentions.*

FOLLOWER EMOTIONAL REACTION
TO THE LEADER'S BEHAVIOR

As Humphrey (2002) has noted, leadership is an emotional process. Dasborough and Ashkanasy (2003) outline leader behaviors that bring about emotional responses in followers. One of the most prominent leader behaviors to evoke positive and negative emotional responses is the leader's attempts to influence followers. In addition, Weiner (1985) has argued that attributions are associated with both influence behavior and the emotional responses to that behavior. While attributions precede emotional responses in Weiner's model (see also Perrewé & Zellars, 1999), followers' attributions and responses were measured simultaneously in the present study, so we make no assumptions about causal direction, testing the relationship as a covariation only. Thus, in our model, positive emotional responses following a leader's influence attempt are likely to covary with attributions of sincere intentions. Similarly, negative emotional responses to the leader's influence attempt are likely to covary with negative attributions of manipulative intentions. Therefore:

Hypothesis 5:   *(a) Followers' positive emotional responses to the leaders' behavior will covary positively with attributions of sincere intentions. (b) Followers' negative emotional responses to the leader's behavior will covary positively with attributions of manipulative intentions.*

METHOD

The experiment is based on a 2 × 2 factorial design (Maxwell & Delaney, 1990). Manipulated independent variables were: (1) leader behavior exhibited, either self-focused or organizationally focused, (2) follower frame of reference, either a target of or a bystander to the leader's behavior. Thus, there were four different treatment conditions, with random assignment of participants. Independent variables were follower mood and work experience. Dependent variables were follower attributions of the leader's sincere and manipulative intentions. Follower positive and negative emotional reactions to the leader's influence attempt were included as covariates of the follower attributions.

Participants

Participants comprised 137 Australian undergraduates studying a course on leadership, 54% of whom were female. They did not receive credit for participating in the experiment; their motivation to participate was to learn about transformational and charismatic leadership. The age of participants ranged from 17 to 45 years, with a mean of 22.8 years (SD = 5.2 years). In terms of place of origin, 63% were Australian (all these Australians were of Anglo origin, none were Australian Aborigines), 23% were Asian, and the remaining 14% were from other parts of the world, mostly

Europe and North America. The mean length of employment was 2.28 years full time and 3.26 years part time.

Materials

Video. The use of videos to portray leadership behaviors is not new to experiments (e.g. Awamleh & Gardner, 1999; Bucey, 2000; Cherulnik, Donley, Wiewel, & Miller, 2000; Glomb & Hulin, 1997; Lewis, 2000). All participants viewed a 3-minute clip of a leader asking them to put in extra effort for the organization. The video was produced by an experienced professional actor, and was based on one segment of a video developed by Awamleh and Gardner (1999). In their study on charismatic leadership, a professional actor was videotaped in four conditions, successfully manipulating content (visionary/nonvisionary) and delivery (strong/weak). For the purpose of this study, the segment depicting the highly charismatic leader was utilized (high vision/strong delivery). In our study, the idea of showing the video was to "put a face" to the leader, to prime participants that the leader is clearly exhibiting transformational and charismatic behavior.

Manipulation stimulus. The manipulation stimuli comprised four versions of an e-mail (printed out), purportedly from the leader shown in the video. All four versions were exactly the same length, and comprised of the same content apart from some minor adjustments. To manipulate self-focused versus organization-focused influence, the wording of the email was "we" and "our organization," compared to the self-focused condition where the words "I" and "my organization" were used. To manipulate frame of reference, the leader's influence attempt was directed to "all people in your Division", in the follower target condition and "all people in another Division" in the bystander condition.

The validity of these manipulations was verified prior to conducting the experiment through pretesting with a sample of postgraduate student judges naive to the study's purpose.

Measures

Attributions of leader intentions. These were measured using a scale based on the Yorges, Weiss, and Strickland's (1990) attribution measure. This scale consists of four items on a five-point scale. We added five new items, which relate specifically to sincere and manipulative intentions. The purpose of these items was to assess pseudo-transformational leader intentions, or true transformational leader intentions. Factor analysis revealed two factors (see Table 10.1 and Results for more details), measuring manipulative intentions (alpha = .72) and sincere intentions (alpha = .66).

Table 10.1.   Factor Analysis Results[a]

Items	Manipulative intention	Sincere intention
Your leader was manipulating you	.68	
Your leader was acting in a self-serving manner	.80	
Your leader was behaving on the basis of beliefs about potential rewards he may gain	.82	
Your leader was behaving on the basis of moral conviction		.68
Your leader was behaving on the basis of his true beliefs		.76
Your leader was acting sincerely		.61
Your leader was behaving on the basis of ethical considerations		.50
Your leader was acting to benefit the organization		.50

Note: [a]All cross loadings < 0.3.

Follower perception of the leader behavior. Follower perceptions of self-focused versus organizationally focused leader behavior were operationalized in terms of ratings on two seven-point semantic differential scales anchored by suitable/unsuitable and appropriate/inappropriate (alpha = .88). High scores indicate perceptions of self-focused behavior.

Follower frame of reference. This was operationalized in terms of the follower's perception of being a bystander or target of the leader's influence attempt. This was measured using a three-item seven-point differential scale anchored, for example, by "intended to engage your effort" versus "not intended to engage your effort" (alpha = 0.73). High scores indicate a perception of the follower as a bystander of the leader's influence attempt.

Follower experience. This was obtained from a self-report measure of work experience. Given that the participants were students and many had only part-time work experience, both full-time and part-time work experience was measured in years.

Emotional reactions. These were measured using Fisher's (2000) Job Emotions Scale (JES). This scale consists of eight negative adjectives and eight positive adjectives assessed on five-point scales. Participants were asked, "How much of each of the following emotions did you feel when you read the e-mail from your leader?" This is the only scale that specifically measures retrospective emotions (not moods) at work. Factor analyses showed that positive emotional responses loaded into one factor (alpha = .85). Negative emotional responses loaded onto a single factor (the highest loading emotional items on this factor were anger, frustration, unhappiness, and disappointment) (alpha = .81).

Current mood. Current mood, or state affect, was measured using the 11-point Faces scale (Fisher, 2000), with 1 = most pleasant or happy (feel great) and 11 = most unpleasant or unhappy (feel bad). Current mood differs from the emotional reactions of the followers, as the mood is not directly in response to an event.

In addition, two categories of variables were measured as potential controls: positive and negative affect (PA and NA) and demographics.

Dispositional affect (trait affect). Trait affect refers to the enduring affective characteristics of a person, or their core affect. This was measured using the Positive and Negative Affect Schedule (PANAS-T) by Watson, Clark, and Tellegen (1988), which measures how members generally feel (alpha reliabilities were .80 for PA and .85 for NA).

Demographic variables. Age, gender, and ethnicity were measured as controls. These control variables, however, were unrelated to any of the variables in the study, so were therefore not included in the data analysis.

Procedure

Data were collected over a semester during regular class meetings. Early in the semester, measures of demographic variables and affective individual differences (trait affect) were gathered from the participants.

The experimental sessions were conducted 10 weeks into the semester. First, the participants completed a measure of current mood. Then, they were told they would be participating in a scenario activity. Participants were told to imagine they were an employee working in an organization under a leader. They were told that this leader will address them as a group via video-conferencing, and then each individual will receive instructions via a personal email. All participants then watched the same video. After the video, each participant received the manipulation e-mail. The four versions of the e-mail were randomly assigned. After reading the e-mail, participants were asked immediately to report their emotional reactions to the leader's behavior, and than completed a questionnaire on their perceptions of the leader's behaviors and their attributions of the leader's sincere and manipulative intentions. Participants were fully debriefed at the end of the experiment to ensure they left informed about the research and in a psychologically well state (Judd, Smith, & Kidder, 1991). Feedback debriefing revealed that no participants were aware of the hypotheses being tested.

RESULTS

Factor Analysis of Attribution Measures

Principal axis factor analysis, followed by Varimax rotation, revealed two factors, based on a scree criterion (see Table 10.1). One represented manipulative intentions (Eigen value = 2.2, 28% variance explained), the other represented sincere intentions (Eigen value = 2.0, 25% variance explained).

Manipulation Checks

Manipulation checks using one-way ANOVAs showed that the two manipulations had been successful. The two experimental groups, self-focused and organizational-focused leader behavior, differed significantly in their perceptions of leader behavior self-focus, self-focused mean = 4.6 (SD = 1.5), organizational-focus mean = 3.8 (SD = 1.5), F (1,134) = 7.9, $p < .01$. The two experimental groups, target and bystander, differed significantly in their perception of being targeted by the leader behavior, target mean = 2.4 (SD = 1.0), bystander mean = 3.1 (SD = 1.5), F (1,135) = 10.6, $p < .01$.

Descriptive Statistics

Descriptive statistics are reported in Table 10.2. These reveal that self-focused leader behavior was positively correlated with attributions of manipulative intentions and negative emotional responses, and negatively correlated with attributions of sincere intentions and positive emotional responses.

Table 10.2. Descriptive Statistics

Variable	Mean[a]	SD	1	2	3	4	5	6	7	8	9	10	11
1. Trait positive affect	3.80	0.49	(.80)[b]										
2. Trait negative affect	2.31	0.68	-.23**	(.85)									
3. Self-focused leader behavior	4.22	1.58	.02	-.10	(.88)								
4. Follower perception as bystander	2.79	1.32	-.09	.13	.18*	(.73)							
5. Full-time work (years)	2.28	5.31	-.03	-.13	.16	-.09							
6. Part-time work (years)	3.26	2.61	.15	-.06	.07	.07	-.26**						
7. Negative mood	4.96	1.93	-.20**	.12	.10	.12	-.17*	-.12					
8. Postive emotional response	1.88	0.83	.11	.00	-.44**	-.22**	-.06	-.20*	-.11	(.85)			
9. Negative emotional response	2.76	1.07	-.00	.04	.49**	.06	.01	.03	.15	-.56**	(.81)		
10. Leader manipulative intentions	3.96	0.79	-.00	.11	.48**	.07	.05	.02	.25**	-.55**	.56**	(.72)	
11. Leader sincere intentions	3.11	0.66	.12	.03	-.49**	-.18**	-.17*	-.08	-.20*	.45**	-.42**	-.43**	(.66)

* p < .05; ** p < .01

Notes: [a] Means are averaged item scores, [b] Alpha reliabilities are shown on the diagonal in parentheses

Hypothesis Tests

Hypotheses 1 and 2 were analyzed using MANOVA. The independent variables were leader behavior (self- vs. organization focus) and follower frame of reference (bystander vs. target). The dependent variables were attributions of manipulative and sincere intentions. Results showed that the interaction was not significant, Wilk's lambda = 0.99, F (2, 132) = 0.98, *ns*. Main effects analyses revealed a significant effect for leader behavior, Wilk's lambda = 0.92, F (2, 133) = 5.70, $p < .01$; but not for frame of reference, Wilk's lambda = 0.99, F (2, 133) = 0.45, *ns*. These results indicated that further analysis of Hypotheses 1a and 1b, relating to leader behavior, were warranted; while Hypotheses 2a and 2b, relating to frame of reference effects, were not supported

Hypotheses 1a and 1b were that followers observing organizationally focused leadership behaviors would be more likely to attribute the behavior to sincere intentions, and to attribute self-focused behaviors' to manipulative intentions. ANOVA results for the main effect revealed that, as expected, participants who received the self-focused e-mails rated manipulative intentions more highly (mean = 4.16, SD = 0.71) than participants who received the organizationally focused e-mails (mean = 3.75, SD = 0.83), F (1,133) = 9.4, $p < .01$. Participants who received the organizationally focused e-mails rated sincere intentions more highly (mean = 3.25, SD = 0.60) than if they had observed self-focused leader behavior (mean = 2.96, SD = .70), F (1,133) = 6.5, $p < .05$.

Multiple regression was used to test Hypothesis 3, which predicted that followers with less work experience will be less likely than more experienced followers to make attributions of manipulative intentions. Results indicated that full-time work experience had no effect on attribution of manipulative intentions, $R^2 < .01$, F (2, 134) = .28, ns. Work experience did, however, predict attributions of sincere intentions, $R^2 = .05$, F (2, 134) = 3.35, $p < .01$. This was because of a significant effect of full-time work, $\beta = -.21$, $p < .05$, suggesting that full-time work experience is associated with less attribution of sincere intention. The beta for part-time work did not reach significance, $\beta = -.14$, *ns*.

Hypotheses 4 and 5 were tested using hierarchical multiple regression (Table 10.3). Hypothesis 4 posited that, when followers are in a positive mood, they are more likely to attribute transformational leader behavior to sincere organizational intentions; and when followers are in a negative mood, they are more likely to attribute transformational leader behavior to manipulative self-focused intentions. To control for trait affect, we entered PA and NA in the first step (Models 1 and 4), then entered current mood. Results (Models 2 and 5) supported Hypothesis 4 for both sincere and manipulative intentions, although the result was only marginally significant

Table 10.3.   Results of Hierarchial Regression Analyses

Variable	Model:	Manipulative intentions			Sincere intentions		
		1	2	3	4	5	6
Trait PA		−.01	.06	.08	.19*	.15	.13
Trait NA		.09	.09	.10	.10	.10	.09
Negative Mood			.26**	.16*		−.16*	−.09
Positive Emotions				−.37**			.30**
Negative Emotions				.33**			−.21*
R^2		.01	.07	.44	.04	.06	.27
Adjusted R^2		−.01	.05	.42	.02	.04	.23
F		.55	3.09*	18.67**	2.27	2.61*	8.47**
ΔR^2			.06	.37		.02	.20
F-Change			8.11**	39.07**		3.21[+]	16.26**

[+] $p < .10$; * $p < .05$; ** $p < .01$
n = 137

in the case of attributions of sincerity. Participants in a bad mood were more likely to make attributions of manipulative intentions, and (marginally) less likely to make attributions of sincere intentions.

Hypothesis 5 was that followers' emotional responses to the leader behavior would covary with their attributions. As shown in Table 10.3 (Models 3 and 6), this hypothesis was strongly supported for both dependent variables. In the case of manipulative intentions, followers reported more negative emotions and less positive emotions; in the case of sincere intentions, followers reported more positive emotional responses and less negative emotional responses. Follower emotions were more strongly related to attributions of leader intentionality than mood. Note, however, that, similar to the results in respect of mood, the effects were stronger in the case of attributions of manipulative intentions (Model 3: $\Delta R^2 = .37$) than for sincere intentions (Model 6: $\Delta R^2 = .20$).

DISCUSSION

The present findings support the propositions of the Dasborough and Ashkanasy (2002) model, that leader behaviors determine follower attributions of leader intentionality. Results reveal in particular that follower mood and emotions are related to follower attributions of intention.

Our results did not, however, support the idea that frame of reference influences attributions of intentionality. In effect, irrespective of whether the leader's influence behavior was directed at the follower's division or another division, the follower made the same attribution. The implication may simply be that observation of the leader's behavior is enough to trigger the attributional response, irrespective of where the leader's influence behavior is directed. On the other hand, this result may be an artifact of the opening video presentation, where the leader is seen to address all members of the organization. In future research, it may be useful to see if the frame of reference manipulation can be strengthened by using different videos for these groups, where the leader is clearly addressing another group in the bystander condition.

Results partially supported the idea that work experience influences attributions of intention. Consistent with the predictions of Ferris and colleagues (1995), we found that years of full-time work were associated with less attribution of sincere intentions. This result may also be interpreted in terms of cynicism, a concept that has recently made its way into management research (Dean, Brandes, & Dharwadkar, 1998). Cynicism has been shown to increase with age (Barefoot, Beckham, Haney, Siegler, & Lipkus, 1993), and this effect has been found in the workplace (Mirvis & Kanter, 1991). This would seem to imply further that leader influence behavior needs to take account of followers' work experience, with possibly more effort needed when trying to influence experienced followers who tend to see such influence behaviors as less sincere.

Some of the most intriguing findings of our research were that follower moods and emotions were so strongly related to attributions of intentionality. Current mood was found to be related to follower attributions of intentionality, especially in the instance of attributions of manipulative intentions. The relationships between emotional reactions and attributions of intentionality, however, were much stronger. This occurred for both positive and negative emotion, and also for attributions of both manipulative and sincere intentions, although the relationship was stronger for manipulative intentions (as for mood effects). Specifically, positive emotions were found to be related to attributions of sincere intentions, while negative emotions were related to attributions of manipulative intentions.

Theoretical Implications

This research makes several contributions to theory development. Leadership research has traditionally focused on cognitive and behavioral aspects of leader behavior (Ashforth & Humphrey, 1995). Recent advances in leadership research have, nonetheless, begun to examine leadership as a distinctly emotional phenomenon (see, e.g., Humphrey, 2002). The findings of the present study indicate that moods and emotions are strongly related to follower attributions about leader intentions. Clearly, if we are to learn more about the complex processes that underlie leadership, future research will need to consider affective as well as cognitive determinants.

Another area of theoretical development concerns the nature of transformational leadership. Our findings suggest that, to evoke positive emotional responses in employees, leaders must display organizationally focused behaviors associated with transformational leadership. Transformational leaders display behaviors associated with individualized consideration, inspirational motivation, intellectual stimulation, and charisma or idealized influence (Bass, 1998). On the other hand, if charismatic leaders display self-focused behaviors, they will be perceived as being manipulative. The findings from this study provide insight into the reasons why some leaders may be labeled by their followers as pseudo-transformational leaders. The key factor influencing this labeling is the attribution of intentionality formed by the follower in response to leadership behavior (Dasborough & Ashkanasy, 2002).

Implications for Research

In a laboratory setting, there is more control over internal validity (Scandura & Williams, 2000), which refers to the question of whether the experimental treatment did indeed have an effect (Mitchell, 1985). Conducting an experiment allowed for the manipulation of leader behavior, so that we could assess the impact of organization-focused versus self-focused leader behavior. Such comparisons in the field are possible, but may be difficult to execute, especially if the leader is truly manipulative and self-serving. Indeed, as Conger (1998) has observed, the study of leadership is highly prone to presentation effects. Often, when asked about leaders, employees will answer in a socially desirable manner. On the other hand, when leaders themselves are asked about their own behaviors, they often attempt to enhance their own image, using various forms of impression management (Conger, 1998).

Practical Implications

The findings of this study also have practical implications. Ashkanasy and Tse (2000), for instance, note that leader–follower relationships can have a profound impact on motivation and work effectiveness. The quality of the leader–follower relationship depends on how followers perceive their leaders, and specifically, their perceptions of the leaders' intentions. Prior research has found that leader–follower relationship quality is related to job performance, satisfaction with supervision, overall satisfaction, commitment, role conflict, role clarity, member competence, and turnover intentions (see, e.g., Gerstner & Day, 1997). Thus, follower attributions of leader intentionality may have considerable impact on the organization.

Other results with practical implications include our finding that frame of reference does not affect attributions of intentionality—although this needs to be tested using a stronger manipulation. Finally, our results suggest that more experienced followers attribute less sincere intentions to their leaders, suggesting that cynicism may play a role with experienced employees.

Limitations and Future Research

There are five potential limitations associated with our study. The first is that our study was conducted in a laboratory setting. Most laboratory experiments, however, do not make claims of external validity (Cook & Campbell, 1976; Mitchell, 1985). These authors argue that participants in laboratory settings may not attribute the same meaning to variables of interest as participants would in field settings (Ilgen, 1986). Locke (1986) and Mook (1983), however, both maintain that results of laboratory studies generalize surprisingly well from the laboratory to the field, especially in psychological research. Furthermore, Martinko and Gardner (1987) have argued specifically that experimental research is needed to understand the underlying mechanism of attribution on organizational settings, while Jung and Avolio (2000) have called for more experimental studies in leadership research.

The second limitation is that there are situations under which people make attributions of leadership, charisma, and effectiveness based on very limited information, for example, first impressions (Awamleh & Gardner, 1999). In this instance, it could be argued that results would be different in real situations, where followers develop relations with their leaders over time. Even if this is so, however, our study suggests that moods and emotions may still play a role. First impressions are an important determinant of the content and outcome of interpersonal encounters (Snyder & Swann, 1978); expectations based on first impressions precipitate expectations for future interpersonal encounters (Harris & Rosenthal, 1985). This is highly relevant in the leader–follower context.

Third, we recognize that use of a student sample may be a weakness, since student samples may not generalize to the broader population (Robson, 1994). Nevertheless, Judd and colleagues (1991) argue that, for most psychological research, college student samples are not functionally different from other samples. In the case of cognitive processes such as attribution formation, we believe that similar findings would occur in the workplace. Also, the students in our sample had at least some work experience. Future research should be conducted in the field to increase realism and to determine if similar results are found in a working sample.

A fourth potential weakness of our study is that we needed to develop a new scale to measure attributions of manipulative and sincere intentions. Although exploratory factor analysis found clear loadings onto the two predetermined factors, the scale for sincere intentions had an alpha reliability if 0.66, marginally below Nunnally's (1978) threshold of .70. Thus, our findings in relation to sincere intentions should be interpreted with caution. Future research should attempt to improve the scales to improve their internal consistency and validity.

Finally, we note that our research does not address the issue of where mood and emotion fit in the attribution process. In the present study, we examined mood as an antecedent to the formation of attributions, and emotion as a covariate. Perrewé and Zellars (1999) argue, however, that emotions can be viewed as an outcome resulting from attributions. Future research using more sophisticated means of investigating causal relationships will be needed to resolve this conundrum.

CONCLUSIONS

The aim of the present experimental study was to examine the determinants of follower attributions of leader intentionality. The two types of attributions followers can make are that the leader behavior is sincere or manipulative. Results revealed that charismatic transformational leaders may be seen as being manipulative if (a) they display self-focused behaviors, (b) the followers are more experienced, and (c) the followers are in a negative mood. We found in addition that (d) followers' experience emotions in response to the leader behavior that are related to their attributions of manipulative versus sincere intentions. Our findings draw attention in particular to follower perceptions of transformational leader behavior, and how moods and emotions may play a critical role alongside these perceptions. As such, we suggest that future examinations of attributions of intentionality should include affective variables, such as trait affect, state affect (current mood), and emotions.

ACKNOWLEDGMENT

This research was funded in part by a grant from the Australian Research Council.

REFERENCES

Ashforth, R. E., & Humphrey, R. H. (1995). Emotion in the workplace: A reappraisal. *Human Relations, 48,* 97–125.

Ashkanasy, N. M. (1995). Supervisory attributions and evaluative judgments of subordinate performance: A further test of the Green and Mitchell model. In M. J. Martinko (Ed.), *Advances in attribution theory: An organizational perspective* (pp. 211–228). Delray Beach, FL: St. Lucie Press.

Ashkanasy, N. M., & Gallois, C. (1994). Leader attributions and evaluations: Effects of locus of control, supervisory control, and task control. *Organizational Behavior and Human Decision Processes, 59,* 27–50.

Ashkanasy, N. M., & Tse, B. (2000). Transformational leadership as management of emotion: A conceptual review. In N. M. Ashkanasy, C. E. J. Härtel, & W. J. Zerbe (Eds.), *Emotions in working life: Theory, research and practice* (pp. 221–235). Westport, CT: Quorum Books.

Awamleh, R., & Gardner, W. L. (1999). Perceptions of leader charisma and effectiveness: the effects of vision, content, delivery and organizational performance. *Leadership Quarterly, 10,* 345–373.

Barefoot, J. C., Beckham, J. C., Haney, T. L., Siegler, I. C., & Lipkus, I. M. (1993). Age-differences in hostility among middle-aged and older adults. *Psychology and Aging, 8,* 3–9.

Baron, R. A. (1989). Impression management by applicants during employment interviews: the "too much of a good thing" effect. In R. W. Eder & G. R. Ferris (Eds.), *The employment interview: Theory, research and practice* (pp. 204–215). Newbury Park, CA: Sage.

Bass, B. M. (1998). *Transformational leadership: Industrial, military, and educational impact.* Mahwah, NJ: Erlbaum.

Bass, B. M., Avolio, B. J., & Atwater, L. (1996). The transformational and transactional leadership of men and women. *Applied Psychology: An International Review, 45,* 5–34.

Betancourt, H., & Blair, I. (1992). A cognition(attribution)-emotion model of violence in conflict situations. *Personality and Social Psychology Bulletin, 18,* 343–350.

Bucy, E. P. (2000). Emotional and evaluative consequences of inappropriate leader displays. *Communication Research, 27,* 194–226.

Cherulnik, P. D., Donley, K. A., Wiewel, T. S. R., & Miller, S. R. (2001). Charisma is contagious: the effect of leaders' charisma on observers' affect. *Journal of Applied Social Psychology, 31,* 2149–2159.

Conger, J. A. (1990). The dark side of leadership. *Organizational Dynamics, 19,* 44–55.

Conger, J. A. (1998). Qualitative research as the cornerstone methodology for understanding leadership. *Leadership Quarterly, 9*, 107–121.

Cook, T. D., & Campbell, D. T. (1976). *Quasi experimentation: Design and analysis issues for field settings*. Boston: Houghton Mifflin.

Dasborough, M. T., & Ashkanasy, N. M. (2002). Emotion and attribution of intentionality in leader-member relationships. *Leadership Quarterly, 13*, 615–634.

Dasborough, M. T., & Ashkanasy, N. M. (2003). Leadership and affective events: How uplifts can ameliorate employee hassles. In C. Cherrey, J. J. Gardner, & N. Huber (Eds.), *Building leadership bridges* (Vol. 3, pp. 58–72). College Park, MD: James MacGregor Burns Academy of Leadership.

Dean, J., Brandes, P., & Dharwadkar, R. (1998). Organizational cynicism. *Academy of Management Review, 23*, 341–352.

Ferris, G. R., Bhawuk, D. P. S., Fedor, D. F., & Judge, T. A. (1995). Organizational politics and citizenship: Attributions of intentionality and construct definition. In M. J. Martinko (Ed.), *Advances in attribution theory: An organizational perspective* (pp. 231–252). Delray Beach, FL: St. Lucie Press.

Fisher, C. D. (2000). Mood and emotions while working: Missing pieces of job satisfaction? *Journal of Organizational Behavior, 21*, 185–202.

Forgas, J. P. (1992). Affect in social judgments and decisions: A multi-process model. In M. Zanna (Ed.), *Advances in experimental social psychology* (Vol. 25, pp. 227–275). San Diego, CA: Academic Press.

Forgas, J. P., & George, J. M. (2001). Affective influences on judgments and behavior in organizations: an information processing perspective. *Organizational Behavior and Human Decision Processes, 86*, 3–34.

George, J. M. (2000). Emotions and leadership: the role of emotional intelligence. *Human Relations, 53*, 1027–1055.

Gerstner, C. R., & Day, D. V. (1997). Meta-analytic review of leader-member exchange theory: correlates and construct issues. *Journal of Applied Psychology, 82*, 827–844.

Glomb, T. M., & Hulin, C. L. (1997). Anger and gender effects in observed supervisor-subordinate dyadic interactions. *Organizational Behavior and Human Decision Processes, 72*, 281–307.

Green, S. G., & Mitchell, T. R. (1979). Attributional processes in leader–member interactions. *Organizational Behavior and Human Performance, 23*, 429–458.

Harris, M. J., & Rosenthal, R. (1985). Mediation of interpersonal expectancy effects: 31 meta-analyses. *Psychological Bulletin, 97*, 363–386.

Heider, F. (1958). *The psychology of interpersonal relations*. New York: Wiley.

Howell, J. M. (1988). Two faces of charisma: socialized and personalized leadership in organizations. In J. Conger & R. Kanungo (Eds.), *Charismatic leadership: The illusive factor in organizational effectiveness* (pp. 213–236). San Francisco: Jossey-Bass.

Humphrey, R. H. (2002). The many faces of emotional leadership. *Leadership Quarterly, 13*, 493–504.

Ilgen, D. R. (1986). Laboratory research: A question of when, not if. In E. A. Locke (Ed.), *Generalizing from laboratory to field settings* (pp. 257–267). Lexington, MA: Lexington Books.

Jones, E .E. (1990). *Interpersonal perception*. New York: Appleton-Century-Crofts.

Judd, C. M., Smith, E. R., & Kidder, L. H. (1991). *Research methods in social relations.* Fort Worth, TX: Holt, Rinehart & Winston.

Jung, D. I., & Avolio, B. J. (2000). Opening the black box: An experimental investigation of the mediating effects of trust and value congruence on transformational and transactional leadership. *Journal of Organizational Behavior, 21,* 949–964.

Leon, I., & Hernandez, J. A. (1998). Testing the role of attribution and appraisal in predicting own and other's emotions. *Cognition and Emotion, 12,* 27–43.

Lewis, K. M. (2000). When leaders display emotion: how followers respond to negative emotional expression of male and female leaders. *Journal of Organizational Behavior, 21,* 221–234.

Locke, E. A. (1986). *Generalizing from laboratory to field settings.* Lexington, MA: Lexingtom Books.

Martinko, M. J., & Gardner, W. L. (1987). The leader member attribution process. *Academy of Management Review, 12,* 235–249.

Maxwell, S. E., & Delaney, H. D. (1990). *Designing experiments and analyzing data: A common comparison perspective.* Belmont, CA: Wadsworth.

Mirvis, P. H., & Kanter, D. L. (1991). Beyond demography—a psychographic profile of the workforce. *Human Resources Management, 30,* 45–68.

Mitchell, T. R. (1985). An evaluation of validity in correlational research conducted in organizations. *Academy of Management Review, 10,* 192–205.

Mitchell, T. R., Green, S. G., & Wood, R. E. (1981). An attributional model of leadership and the poor-performing subordinate: Development and validation. *Research in Organizational Behavior, 3,* 197–234.

Mook, D. M. (1983). In defense of external validity. *American Psychologist, 38,* 379–387.

Nunnally, J. C. (1978). *Psychometric theory* (2nd ed.). New York: McGraw-Hill.

Owen, H. (1986). Leadership indirection. In J. D. Adams (Ed.), *Transformational leadership* (pp. 111–122). Alexandria, VA: Miles River Press.

Perrewé, P. L., & Zellars, K. L. (1999). An examination of attributions and emotions in the transactional approach to the organizational stress process. *Journal of Organizational Behavior, 20,* 739–752.

Robson, C. (1994). *Experiment, design and statistics in psychology.* Harmondsworth, UK: Penguin Education.

Scandura, T. A., & Williams, E. A. (2000). Research methodology in management: current practices, trends, and implications for future research. *Academy of Management Journal, 43,* 1248–1264.

Smith, C. A., Haynes, K. N., Lazarus, R. S., & Pope, L. K. (1993). In search of "hot" cognitions: Attributions, appraisals, and their relation to emotion. *Journal of Personality and Social Psychology, 65,* 916–929.

Snyder, M., & Swann, W. B. (1978). Behavioral confirmation in social interaction: From social perception to social reality. *Journal of Experimental Social Psychology, 14,* 148–162.

Watson, D., Clark, L. A., & Tellegen, A. (1988). Development of brief measures of positive and negative affect: the PANAS scale. *Journal of Personality and Social Psychology, 54,* 1063–1070.

Weierter, S. J. M. (1997). Who wants to play "follow the leader?" A theory of charismatic relationships based on routinized charisma and follower characteristics. *Leadership Quarterly, 8,* 171–193.

Weiner, B. (1977). Attribution and affect: Comments on Sohn's critique. *Journal of Educational Psychology, 69,* 506–511.

Weiner, B. (1985). An attributional theory of achievement motivation and emotion. *Psychological Review, 92,* 548–573.

Weiner, B., Graham, S., & Chandler, C. (1982). Pity, anger, and guilt: An attributional analysis. *Personality and Social Psychology Bulletin, 8,* 226–232.

Yorges, S. L., Strickland, O. J., & Weiss, H. M. (1999). The effect of leader outcomes on influence, attributions, and perceptions of charisma. *Journal of Applied Psychology, 84,* 428–436.

CHAPTER 11

CONFLICT MANAGEMENT

An Attributional Perspective

Charles Joseph and Scott Douglas
Binghamton University

ABSTRACT

People engage in one of five conflict resolution behaviors when faced with a conflict situation. These behaviors are often described as a function of two underlying dimensions, concern for self and concern for others. However, little is known about the factors that produce these concerns. In this chapter we develop a model that shows that concern for self and concern for others are influenced by attributions as to why there is a specific conflict and attribution styles as to why conflicts occur in general.

Interpersonal conflict, a process in which one party perceives that another party is blocking it from desired goals or interests (Rahim, Magner, & Shapiro, 2000; Wall & Callister, 1995), is pervasive in the workplace (Kolb & Putnam, 1992; Pondy, 1967; Putnam, 1994; Van de Vliert, 1997). This conflict can result in positive (e.g., reaching better decisions and greater employee satisfaction, Korbanik, Baril, & Watson, 1993; Tutzauer & Roloff, 1988; increased team and organizational performance, Likert & Likert, 1976; Vigil-King, 2000) and negative outcomes (e.g., increased employee hostility, anger, and deviance, Aquino & Douglas, 2003; Douglas & Mar-

Attribution Theory in the Organizational Sciences, pages 225–242
Copyright © 2003 by Information Age Publishing
All rights of reproduction in any form reserved.

tinko, 2001; Thomas, 1976). Consequently, the constructive management of interpersonal conflict remains central to the effectiveness of organizations (Rahim, 1997; 2000). Moreover, while individuals who are perceived as effective conflict managers are also seen as capable leaders (Gross & Guerrero, 2000), those who are viewed as ineffective conflict managers are likely to experience difficulties in maintaining positive relationships (Canary, Cupach, & Messman, 1995) and reaching organizational goals (Nicotera, 1995).

Given the inevitability for experiencing conflict in the workplace (Putnam, 1994) and its potential to impact organizational performance (positively and negatively), there is growing interest in the academic community toward understanding the factors that influence conflict resolution behaviors (Rahim et al., 2000). This attention has stimulated an interesting and ongoing debate about whether people choose a particular conflict resolution behavior based on situational factors associated with a specific conflict (i.e., conflict resolution strategy, Drory & Ritov, 1997; Knapp, Putnam, & Davis, 1988) or exhibit a disposition toward a particular resolution behavior that is relatively stable across conflict settings (i.e., conflict resolution style, Rahim, 1992; Sternberg & Dobson, 1987). It is our contention, however, that this debate points toward the lack of a theoretical framework that allows for situational and dispositional influences on the choice of conflict resolution behavior. Furthermore, whereas the situational perspective de-emphasizes the likelihood that people may rely on personal preferences for dealing with conflict situations, particularly when these events are ambiguous, the dispositional approach minimizes the potential for peoples' thought about the causes that manifest a particular conflict to influence their strategic response. In light of these shortcomings, we propose and develop an attributional model illustrating that the behaviors people exhibit when dealing with a conflict situation are dependent upon their attributions (Weiner, 1985, 1995) as to why there is a specific conflict, which may be indicative of situational influences and peoples' attribution styles (Abramson, Seligman, & Teasdale, 1978; Kent & Martinko, 1995; Martinko, 2002; Russell, 1991) as to why conflicts occur in general, which is suggestive of dispositional influences.

The literature on conflict management often focuses on five, previously identified resolution behaviors (i.e., avoiding, obliging, dominating, integrating, and compromising, Rahim & Bonoma, 1979), which are argued to be a function of two underlying dimensions, concern for self and concern for others. However, this literature does not explicitly address why someone is likely to show more or less concern for self or others. Without such understanding, the conflict literature is likely to be perceived as largely descriptive. An attributional perspective on conflict however, provides a descriptive and prescriptive framework. It is descriptive to the extent it explains the reasons why people exhibit different levels of concern for self and others and thus engage in different conflict resolution behaviors, and prescriptive to the extent it suggests remedies such as attributional training (Albert, 1983; Fiedler, Mitchell, & Triandis, 1971; Martinko, 1995; Martinko & Gardner, 1982) and attributional cueing (Lee, Hallahan, & Herzog, 1996) that can be implemented to promote more effective conflict resolution in organizations.

The situational approach to conflict management (e.g., Drory & Ritov, 1997; Knapp et al., 1988; Thomas, 1979) emphasizes elements of the situation (e.g., differences in power between conflicting parties, time pressures, significance of goals) that motivate conflict resolution behaviors. Yet the attribution literature (e.g., Martinko, 1995, 2002; Weiner, 1985, 1995) clearly indicates that an individual's causal explanation for an event, particularly negative events such as many conflict episodes, can also motivate a behavioral reaction (Gundlach, Douglas, & Martinko, 2003), including those exhibited in conflict situations. Additionally, the literature on attribution styles (e.g., Martinko, 2002) suggests that some people will demonstrate an enduring tendency to use a particular resolution behavior—a conflict resolution style—regardless of its appropriateness for a given situation.

In the following pages, we briefly review the literature on conflict management, paying particular attention to the debate on situational and dispositional determinants of conflict resolution behavior. Afterward, we recap the attribution literature, including a discussion of attribution styles. Based on these discussions, we develop our model illustrating how peoples' attributions and attribution styles motivate different conflict resolution behaviors via their impact on concern for self and concern for others. Then, we discuss the theoretical and practical implications of our proposed framework, highlight some opportunities for future investigations, acknowledge limitations of the proposed model, and make concluding comments. Accordingly, it is to the conflict management literature that we now turn our attention.

CONFLICT MANAGEMENT

Workplace conflict, which is a frequent outcome of interpersonal interactions among employees, begins when one party believes that its goals or interests are obstructed by another party (Rahim et al., 2000; Wall & Callister, 1995). The presence of conflict in organizations can have both positive and negative consequences. Positive consequences can include enhanced creativity and innovation, higher quality decision making, and improved mutual understanding between organizational members (De Dreu, 1997;

Pelted, Eisenhardt, & Xin, 1999). However, perhaps the most evident outcomes of conflict situations are the negative consequences, such as increased employee anxiety, anger and hostility (Ephross & Vassil, 1993; Thomas, 1976), social and emotional separation from the workplace (Retzinger, 1991) and several counterproductive forms of behavior such as decreased work effort (Greenberg, 1999), and workplace aggression and sabotage (Martinko, Gundlach, & Douglas, 2002).

Constructive management of organizational conflict requires that its negative consequences be minimized and its positive consequences maximized. Research indicates that leaders who are not adept at handling conflict experience greater difficulty in reaching organizational goals (Nicotera, 1995). Moreover, according to Infante, Anderson, Martin, Herington, and Kim (1993), these individuals are also more likely to express decreased job satisfaction. In contrast, supervisors who demonstrate effective conflict resolution behaviors are often seen as competent, effective leaders (Gross & Guerrero, 2000). Given these outcomes, it is not surprising that significant attention has been placed on identifying conflict resolution behaviors and the factors that influence their use.

Over 60 years ago, Mary Parker Follett (1940) argued that there are three primary (i.e., domination, integration, and compromise) and two secondary (i.e., avoidance and suppression) conflict resolution behaviors. Since then, researchers have conceptualized these behaviors as falling along two underlying dimensions. Blake and Mouton (1964) classified the behaviors along the dimensions of concern for production and concern for people. Later, Rahim and his colleagues (Rahim 1983, 1985, 1992, 2000, 2001; Rahim & Bonoma, 1979; Rahim & Magner, 1995) indicated that while there are five prominent behaviors for conflict resolution, including dominating (i.e., a win–lose orientation), integrating (i.e., a win–win orientation), avoiding (i.e., a side-stepping orientation), obliging (i.e., a lose–win orientation), and compromising (i.e., a give-and-take orientation), these behaviors were better conceived as reflecting the two underlying dimensions of concern for self and concern for others. Support for their contention regarding the two underlying dimensions was yielded in a later study conducted by Van de Vilert and Kabanoff (1990).

While there appears to be growing recognition and acceptance of the five conflict resolution behaviors and the underlying dimensions of concern for self and concern for others, there is some debate as to whether the use of these strategies is a function of the conflict situation or an underlying predisposition toward a particular behavior (Freidman, Tidd, Currall, & Tsai, 2000). Moreover, advocates of the situational perspective (e.g., Knapp et al., 1988; Lewicki & Shepherd, 1985; Pruitt, 1983; Shepherd, 1984; Thomas, 1979) argue that the five behaviors should not be viewed as traits or styles since there is some evidence (e.g., Drory & Ritov, 1999) that people match their resolution behaviors to particular circumstances. Nevertheless, the belief that situational characteristics can influence one's conflict resolution behavior does not preclude dispositional influences on these behaviors (Freidman et al., 2000)

Conflict resolution style is an enduring tendency to use a particular resolution behavior (Gross & Guerrero, 2000), regardless of its appropriateness for addressing a given conflict situation. For example, some people may exhibit a strong preference to use a dominant approach, while for the same conflict situation others exhibit an equally strong preference toward an obliging approach. Along this line of explanation, advocates of a dispositional approach (e.g., Snyder & Ickes, 1985; Sternberg & Dobson, 1987) to conflict management suggest that the reason people exhibit these different tendencies is because of differences in underlying dispositions that manifest higher or lower concern for self and concern for others in general.

In summary, there seems to be support for the notion that elements of the situation and person influence the choice of a conflict resolution behavior. And, although there is growing evidence that the presence of significant situational influences do not necessarily preclude individual differences from impacting conflict resolution behaviors (e.g., Graziano, Jenson-Campbell, & Hair, 1996; Sternberg & Soriano, 1984), the field appears to lack a conceptual framework that allows for this coexistence. Furthermore, the research on conflict management has yet to consider the implications of an individual's causal attributions as to why they are faced with a conflict or why conflicts occur in general. Certainly, there is substantial evidence that one's causal analyses of events, particularly negative events (e.g., conflict situations), influence their subsequent reactions. It is to this body of evidence and related theory that we now turn.

ATTRIBUTIONS

Attribution theory, which is in part based on the metaphor of individuals as naïve psychologists (Heider, 1958), proposes that people attempt to infer causes for observed behavior. This body of theory is also concerned with the consequences of these perceptions (Martinko, 1995) since, as described by Weiner (1986, 1995), attributions can play a central role in motivated behavior.

Attributional explanations can often be categorized along two underlying dimensions (i.e., locus of causality and stability), which have been related to affective reactions, expectancies, and subsequent behaviors (Martinko & Gardner, 1987). Along the locus of causality dimension, internal attributions for failure such as to a perceived lack of ability or insufficient effort can manifest negative, self-directed emotions and behaviors like shame, guilt, learned-helplessness, and passivity (Campbell & Martinko, 1998; Martinko & Gardner, 1982; Martinko et al., 2002). External attributions for such events, however, can produce emotions of anger and frustration (Aquino, Douglas, & Martinko, in press) as well as aggressive behaviors (Dodge & Coie, 1987; Martinko et al., 2002; Martinko & Zellars, 1998), which can be viewed as self-protecting (Aquino & Douglas, 2003). The stability of a perceived cause, though, impacts on individuals' expectancies for change in the future (Martinko & Gardner, 1987). To the extent people believe that the cause for an event is stable, they are less likely to see the potential for change and thus more likely to expect similar outcomes in the future.

Though attribution theorists are concerned with the consequences of causal perceptions, attributions themselves may not be reflective of the actual causes for an event. Moreover, research indicates that people can exhibit attribution or causal reasoning styles, which refer to consistent patterns of making attributions for one's outcomes to similar causal dimensions, regardless of a particular situation (Abramson et al., 1978; Anderson, Jennings, & Arnoult, 1988; Martinko, 2002). Consequently, people are more likely to make attributions that fail to reflect the actual causes for a particular event to the extent they exhibit an attribution style.

Attribution style is an individual difference variable that has been shown to predict several intra- and interpersonal phenomena. The results of several studies indicate that attribution styles are associated with overt expressions of anger (Aquino et al., in press), aggression in young males (Dodge & Coie, 1987; Nasby, Haden, & DePaulo, 1979), propensity to engage in acts of workplace aggression (Douglas & Martinko, 2001), learned helplessness (Abramson et al., 1978), quality of relations in leader–member dyads (Martinko & Moss, 2000), and the successful adoption of new technologies (Henry, Martinko, & Pierce, 1993).

Arguably, the literature has emphasized the influence of intrapersonal (i.e., attributions made for our own outcomes) and interpersonal attributions (i.e., attributions we make for the outcomes of others) on individuals' affective reactions, expectancies, and social exchanges. This literature has also begun to recognize the role that intrapersonal attribution styles can play in distorting causal perceptions, which can produce a variety of negative outcomes. Yet little attention has been given to the role interpersonal (social) attribution styles can play in organizational settings (Martinko, Moss, & Douglas, 2003).

Recently, Martinko (2002) suggested that, because people often think differently about their own outcomes than they do about the outcomes of others, individuals could also exhibit interpersonal attribution styles (i.e., cross-situational tendencies to attribute the outcomes of others to similar causal dimensions). More importantly, though, similar to intrapersonal attribution styles, interpersonal attribution styles are likely to play an influential role in how people behave toward others over time (Martinko et al., 2003). This is a particularly relevant observation in the context of conflict management given that it describes a situation wherein one party is likely to experience a negative event in which another party is involved. Therefore, in the next section we move on to develop our model and related propositions describing the relations between intra- and interpersonal attributions and attribution styles, concern for self and others, and conflict resolution behaviors.

MODEL AND PROPOSITION DEVELOPMENT

The Attributional Model of Conflict Resolution Behaviors (Figure 11.1) illustrates that concern for self and others are influenced directly and indirectly by attributions and attribution styles. Thus, concern for self and concern for others are reflective of our intra- and interpersonal attributions and attribution styles. Our model also shows that while the locus of causality dimension is the primary determinant of concern for self and concern for others, the stability dimension is the primary determinant of whether a person is more or less likely to engage in compromising behaviors as opposed to one of the other four resolution behaviors.

The rationale for our model is described below. Before we engage in this discussion, however, we first recognize a significant assumption underpinning the model as well as identify an important constraint. Although conflict is described as an event in which one party's (A) goal is obstructed by another party (B), we assume that people can make different attributions as to why the other party is blocking them. For instance, party A could attribute party B's blocking behavior to external sources (e.g., we could not negotiate an agreement because party B was following a company policy that was inconsistent with our demands), which is likely to produce a different reaction than if the blocking behavior is attributed to internal causes (e.g., we could not negotiate an agreement because party B is an egomanic who has to win at all costs). Regarding the constraint, the present discussion is limited to conflict situations that result from negative events. This restriction is due to a need to limit the scope and length of our present discussion and the observation that people are more likely to engage in attribution processes following negative rather than positive events (Wong & Weiner, 1981). Nonetheless, we recognize that conflict can arise from positive situations such as when a work team has received a monetary reward for its performance and has to determine how the award will

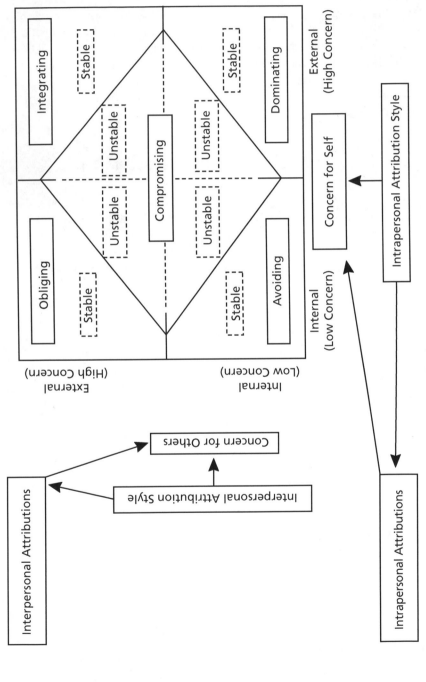

Figure 11.1. Attributional Model of Conflict Resolution Behaviors

be distributed amongst its members. All too often team members form different reasons for their team's success that in turn manifest conflicts about what each member deserves, or does not deserve.

Results of both conceptual (e.g., Martinko et al., 2002) and empirical works (e.g., Perlow & Latham, 1993) suggest that when people make intrapersonal attributions in which they attribute negative events to external causes they are more likely to engage in self-protecting behaviors, which can be viewed as more self-concerning and less other-concerning behaviors. For instance, Aquino and colleagues (in press) argued and demonstrated that when people attribute negative occurrences to external causes they are more likely to express their anger toward others as a means for reaffirming their self-image as someone deserving of respect. The results of a study conducted by Perlow and Latham (1993) also suggest that, following negative events, individuals with an external locus of control act abusively toward others. Both examples are indicative of someone who demonstrates high concern for self and low concern for others following negative events. Moreover, based on Rahim and his colleagues' work (e.g., Rahim 1983, 1985, 1992, 2000, 2001; Rahim & Bonoma, 1979; Rahim & Magner, 1995) these individuals should prefer a dominating (high concern for self, low concern for others) conflict resolution behavior wherein their objective is to win, often at the expense of the other party.

Research on self-esteem suggests that people tend to engage in passive behaviors that reflect a lower self-concern when they make internal intrapersonal attributions for failure. In contrast, those who are high in self-esteem tend to make external attributions for their negative outcomes (Levy, 1993) and tend to be oriented toward self-enhancement (Baumeister, 1995). People with low self-esteem, however, tend to make internal attributions for failure (Levy, 1993), and are oriented toward self-protection and avoidance behaviors (Baumeister, 1995). Therefore, based on these observations and our previous arguments, we anticipate that individuals who make internal intrapersonal attributions for negative outcomes would demonstrate low concern for self and high concern for others and engage in obliging behaviors; or demonstrate low concern for self and low concern for others and engage in avoiding behaviors since there is an element of self-sacrifice or withdrawal in both.

When we turn to interpersonal attributions, there is evidence to suggest that people exhibit lower (higher) concern for others when they make internal (external) attributions for others' failures. The justice literature indicates that people are more sensitive and responsive to a perceived injustice when they attribute the injustice to the characteristics of the perpetrator rather than external factors (e.g., Cohen, 1982; Utne & Kidd, 1980). Furthermore, the just-world hypothesis (Lerner, 1980) suggests that when people make interpersonal attributions indicating that the cause for a negative event is due to internal factors, they view the other person as deserving whatever they get. Yet they view the other person as a victim who should be helped when they make interpersonal attributions indicating external causes for negative events. Thus, in the first case they should show lower concern for others and engage in either avoiding or dominating behaviors, while in the second case they should demonstrate higher concern for others and engage in either obliging or integrating behaviors.

Whereas dominating, avoiding, obliging, and integrating conflict resolution behaviors are largely influenced by locus of causality perceptions, the potential for someone to use a compromising behavior is largely influenced by perceptions of stability. Compromising involves a give-and-take relationship wherein both parties to a conflict are willing to give up something to achieve a mutually acceptable outcome (Rahim, 1985). Inherent in this description is the need for flexibility in the future. Hence, a person who perceives the possibility for change on the part of both parties to the conflict is more likely to engage in compromising behavior than a person who believes the underlying causes for the conflict are stable (at least in reference to one of the two parties) and therefore unlikely to change. For example, party A is more likely to engage in compromising behavior when it believes that a conflict situation is due to its lack of preparedness (unstable) for negotiating an agreement and party B being misinformed (unstable) about some of the issues related to the negotiation than if party A perceives that the conflict is due to party B's unscrupulous nature (stable) and circumstances of chance for party A (unstable). Moreover, we would anticipate that in the latter case, party A would engage in a dominating approach since there is no foreseeable potential for change on at least one of the sides; and party A has made external intrapersonal attributions that stimulate higher concern for self, while making internal interpersonal attributions that stimulate lower concern for others.

Combining the previous arguments on locus of causality, concern for self and concern for others, and stability results in 16 potential combinations for making intra- and interpersonal attributions for conflict situations. Each combination of attributional explanations leads to a specific prediction for level of concern for self and concern for others and therefore the use of a particular conflict resolution behavior. However, rather than formally stating 16 separate propositions, we have chosen to illustrate them in Table 11.1, which, for example, indicates that people will engage in avoiding behaviors rather than dominating, obliging, integrating or compromising behaviors when they make internal and stable intrapersonal attributions (e.g., "I am not a good negotiator") and internal and stable interpersonal attributions (e.g., "The other party does not care about our interests"). Lastly, because attribution styles predispose people to make attributions to similar causal dimensions over time and situations (Anderson et al., 1988), particularly when the conflict situations are ambiguous

Table 11.1.

	Intrapersonal Attributions			Interpersonal Attributions			
Proposition	LOC	Stability	Self-Concern	LOC	Stability	Other-Concern	Resolution Behavior
1	External	Stable	High	Internal	Stable	Low	Dominating
2	External	Stable	High	Internal	Unstable	Low	Dominating
3	External	Unstable	High	Internal	Stable	Low	Dominating
4	External	Unstable	High	Internal	Unstable	Low	Compromising
5	Internal	Stable	Low	Internal	Stable	Low	Avoiding
6	Internal	Stable	Low	Internal	Unstable	Low	Avoiding
7	Internal	Unstable	Low	Internal	Stable	Low	Avoiding
8	Internal	Unstable	Low	Internal	Unstable	Low	Compromising
9	Internal	Stable	Low	External	Stable	High	Obliging
10	Internal	Stable	Low	External	Unstable	High	Obliging
11	Internal	Unstable	Low	External	Stable	High	Obliging
12	Internal	Unstable	Low	External	Unstable	High	Compromising
13	External	Stable	High	External	Stable	High	Integrating
14	External	Stable	High	External	Unstable	High	Integrating
15	External	Unstable	High	External	Stable	High	Integrating
16	External	Unstable	High	External	Unstable	High	Compromising

Note: LOC = locus of causality

(Snyder & Ickes, 1985), we predict that attribution styles will manifest particular conflict resolution styles over time that are also consistent with those depicted in Table 11.1.

DISCUSSION

Organizations are more effective when functional behaviors are maximized and dysfunctional behaviors are minimized. The ability to successfully manage conflict is therefore paramount to organizational success. At the individual level, conflict management skills are important determinants of one's ability to work effectively with others on a continuous basis. Consequently, a clear understanding of the factors that influence conflict management behaviors is crucial to the continued success of organizations and their members.

In this chapter we used attribution theory to illustrate how conflict resolution behaviors can be motivated by peoples' perceptions of the causes for a particular conflict situation as well as their perceptions of why conflicts occur in general. In so doing, we contend that our application of attribution theory to the area of conflict management further demonstrates the theory's utility for integrating the environmental and individual influences on behavior. Admittedly, though, we recognize that causal perceptions of situational attributes may not depict the actual causes for a particular conflict episode. Nonetheless, this observation does not undermine the importance for understanding that conflict resolution behaviors are motivated behaviors, which in part are determined by causal perceptions.

Freidman and colleagues (2000) have shown that people who exhibit a particular conflict resolution style can create environments with varying degrees of conflict. People who demonstrate an integrating style produce environments with less conflict relative to those who exhibit dominating or avoiding styles (Freidman et al., 2000). To the extent this is correct and our model is valid, we would anticipate that those people who demonstrate strong attributional styles would play a significant role in determining workplace cultures. Conceptually this is important since it points to the possibility that several of the negative and positive outcomes that have been associated with various types and levels of organizational culture may be better understood if they were examined through an attributional-cultural lens.

Practitioners who are adept at using several of the conflict resolution behaviors have a significant advantage over those with limited conflict-management skills. Hence, the importance of educating organizational members on conflict resolution behaviors cannot be understated. However, the efficacy of such programs is likely to be undermined if practitioners fail to recognize that these are in part motivated behaviors, which are likely to be influenced on a case-by-case basis by causal perceptions. In addition, the value of conflict management seminars is further suspect to the extent the participants exhibit attributional styles that motivate behaviors that are inconsistent with the espoused behaviors. The presence of strong role norms and reward systems may produce more appropriate conflict resolution behaviors (Trevino & Victor, 1992), but these prevention strategies are likely to be insufficient to the extent that attribution styles are trait-like. In light of this potential, practitioners may implement programs such as attributional training and attributional cueing. Attributional training (Albert, 1983; Fiedler et. al., 1971; Martinko & Gardner, 1982) places an emphasis on making people aware of their attributional biases and how these biases influence their social interactions. Attributional cueing also focuses on making people aware of their attributional biases; however, this approach is aimed at getting people to engage in higher levels of cognitive complexity, which has been shown to reduce attributional biases (Lee et. al., 1996).

Making accurate attributions does not, however, assure the use of an appropriate resolution behavior for a given conflict situation. Moreover, it is quite plausible that accurate attributions have the potential to manifest a conflict resolution behavior that is most undesirable at that time. Therefore, practitioners should avoid the temptation to implement only one set of prevention programs, but instead recognize the value of providing programs that educate members on both attributions and conflict resolution behaviors.

Obviously, it is our desire that our model be tested. Given that the literature suggests that an integrating style is viewed as a more effective approach to managing conflict than dominating and avoiding styles (Burke, 1970), we anticipate that the attributional combinations shown in Table 11.1 could also be used to predict perceptions of management effectiveness. Hence, this line of research may be beneficial in the future as we try to gain a greater understanding of the determinants of effective management.

Although our focus was limited to the attributional implications of conflict management, we recognize that contextual factors such as differences in status and/or power between conflicting parties, organizational norms (Snyder & Ickes, 1985), and conflict type (i.e., interpersonal vs. interorganizational, Sternberg & Soriano, 1984) may influence the relations depicted by our model. We do suspect, though, that contextual factors such as these will moderate the relations between attributions, attribution styles, and conflict resolution behaviors. Therefore, future research along this line is also encouraged.

Another potential limitation is that this study examined perceptions stemming only from negative events. There may be important differences in how people react based on whether the conflict stems from negative or positive outcomes. Such a possibility is consistent with Martinko's (2002) recent discussion in which he argues that people can demonstrate different attributional tendencies based on whether the related events are positive or negative. Consequently, researchers may consider how the relations between conflict resolution behaviors, attributions, and attribution styles differ as a function of how positively the conflict situation is viewed.

CONCLUSION

The need to manage conflict in organizations is likely to rise as we enter a more global marketplace in which issues of diversity and cultural misunderstandings become more prevalent. The model we developed extends our theoretical understanding of this phenomenon, and it is our hope that it sheds additional light on intervention strategies that can enhance practitioners' ability to manage conflict effectively.

REFERENCES

Abramson, L. Y., Seligman, M. E. P., & Teasdale, J. D. (1978). Learned-helplessness in humans: Critique and reformulation. *Journal of Abnormal Psychology, 87*, 49–74.

Albert, R. (1983). The intercultural sensitizer of cultural assimilator: A cognitive approach. In D. Landis & R. Brislin (Eds.), *Handbook of intercultural training* (Vol. 2, pp. 186–217). New York: Pergamon.

Anderson, C. R., Jennings, D., & Arnoult, L. (1988). Validity and utility of the attributional style construct at a moderate level of specificity. *Journal of Personality and Social Psychology, 55*, 979–990.

Aquino, K., & Douglas, S. C. (2003). Identity threat and antisocial behavior in organizations: The moderating effects of individual differences, aggressive modeling and hierarchical status. *Organizational Behavior and Human Decision Processes, 90*, 195–208.

Aquino, K., Douglas, S. C., & Martinko, M. J. (in press). Overt expressions of anger in response to perceived victimization: The moderating effects of attributional style, hierarchical status, and organizational norms. *Journal of Occupational Health Psychology.*

Baumeister, R. F. (1995). Self and identity: An introduction. In A. Tesser (Ed.), *Advanced social psychology* (pp. 50–97). New York: McGraw-Hill.

Blake, R. R., & Mouton, J. S. (1964). *The managerial grid.* Houston, TX: Gulf.

Burke, R. J. (1970). Methods of resolving superior-subordinate conflict: The constructive use of subordinate differences and disagreements. *Organizational Behavior and Human Performance, 5*, 393–411.

Campbell, C. R., Martinko, M. J. (1998). An integrative attributional perspective of empowerment and learned-helplessness: A multi-method field study. *Journal of Management, 24*, 173–200.

Canary, D. J., Cupach, W. R., & Messman (1995). *Relational conflict: Conflict in parent–child, friendship, and romantic relationships.* Thousand Oaks, CA: Sage.

Cohen, R. L. (1982). Perceiving injustice: An attributional perspective. In J. Greenberg & R. L.

Cohen (Eds.), *Equity and justice in social behavior* (pp. 119–160). New York: Academic Press.

De Dreu, C. K. W. (1997). Productive conflict: The importance of conflict management and conflict issue. In C. K. W. De Dreu & E. Van De Vliert (Eds.), *Using conflict in organizations* (pp. 9–22). London: Sage.

Dodge, K. A., & Coie, J. D. (1987). Social information processing factors in reactive and proactive aggression in children's peer groups. *Journal of Personality and Social Psychology, 53*, 1146–1158.

Drory, A., & Ritov, I. (1997). Effects of work experience and opponent's power on conflict management styles. *International Journal of Conflict Management, 8,* 148–161.

Douglas, S. C., & Martinko, M. J. (2001). Exploring the role of individual differences in the prediction of workplace aggression. *Journal of Applied Psychology, 86,* 547–559.

Ephross, R. H., & T. V. Vassil (Eds.). (1993). *Social work with groups: Expanding horizons.* Binghamton, NY: Haworth Press.

Fiedler, F., Mitchell, T., & Traindis, H. (1971). The cultural assimilator: An approach to cross-cultural training. *Journal of Applied Psychology, 55,* 95–102.

Follett, M. P. (1940). Constructive conflict. In H. C. Metcalf & L. Urwick (Eds.), *Dynamic administration: The collected papers of Mary Parker Follett* (pp 30–49). New York: Harper. (Original work published 1926).

Friedman R. A., Tidd, S. T., Currall, S. C., & Tsai, J. C. (2000). What goes around comes around: The impact of personal conflict style on work conflict and stress. *International Journal of Conflict Management.*

Graziano, W. G., Jensen-Campbell, L. A., & Hair, E. C. (1996). Perceiving interpersonal conflict and reacting to it: The case for agreeableness. *Journal of Personality and Social Psychology, 70,* 820–835.

Greenberg, J. (1999). *Managing behavior in organizations* (2nd ed.). Upper Saddle River, NJ: Prentice-Hall.

Gross, M. A., & Guerrero, L. K. (2000). Managing conflict appropriately and effectively: An application of the competence model to Rahim's organizational conflict strategies. *International Journal of Conflict Management.*

Gundlach, M. J., Douglas, S. C., & Martinko, M. J. (2003). The decision to blow the whistle: A social information processing framework. *Academy of Management Review, 28,* 107–123.

Heider, F. (1958) *The psychology of interpersonal relations.* New York: Wiley.

Henry, J. W., Martinko, M. J., & Pierce, M. A. (1993). Attributional style as a predictor of success in a first computer science course. *Computers in Human Behavior, 342–352.*

Infante, D. A., Anderson, C. M., Martin, M. M., Herington, A. D., & Kim, J. (1993). Subordinates' satisfaction and perceptions of superiors' compliance-gaining tactics, argumentativeness, verbal aggressiveness, and style. *Management Communication Quarterly, 6,* 307–326.

Kelley, H. H. (1973). The process of causal attributions. *American Psychologist, 28,* 107–128.

Kent, R., & Martinko, M. J. (1995). The development and evaluation of a scale to measure organizational attribution style. In M. J. Martinko (Ed.), *Attribution theory: An organizational perspective* (pp. 53–75). Delray Beach, FL: St. Lucie Press.

Knapp, M. L., Putnam, L. L., & Davis, L. J. (1988). Measuring interpersonal conflict in organizations: Where do we go from here? *Management Communication Quarterly, 1,* 414–429.

Kolb, D. M., & Putnam, L. L. (1992). The multiple faces of conflict in organizations. *Journal of Organizational Behavior 13,* 311–324.

Korbanik, K., Baril, G. L., & Watson, C. (1993). Managers' conflict management style and leadership effectiveness: The moderating effects of gender. *Sex Roles, 29*, 405–420.

Lee, F., Hallahan, M., & Herzog, T. (1996). Explaining real-life events: How culture and domain shape attributions. *Personality and Social Psychology Bulletin, 22*: 732–741.

Lerner, M. J. (1980). *The belief in a just world: A fundamental delusion.* New York: Plenum Press.

Levy, P. E. (1993). Self-appraisal and attributions: A test of a model. *Journal of Management, 19*, 51–63.

Lewicki, R., & Shepherd, B. (1985). Choosing how to intervene: Factors influencing the use of process and outcome control in third party dispute resolution. *Journal of Occupational Behavior, 6*, 49–64.

Likert, R., & Likert, J. G. (1976). *New ways of managing conflict.* New York: McGraw-Hill.

Martinko, M.J. (Ed.). (1995). *Attribution theory: An organizational perspective.* Delray Beach, FL: St. Lucie Press.

Martinko, M. J. (2002). *Thinking like a winner: A guide to high performance leadership.* Tallahassee, FL: Gulf Coast.

Martinko, M. J., & Gardner, W. L. (1982). Learned-helplessness: An alternative explanation for performance deficits. *Academy of Management Review, 7,* 195–204.

Martinko, M. J., & Gardner, W. L. (1987). The leader-member attribution process. *Academy of Management Review, 12*, 235–239.

Martinko, M. J., Gundlach, M. J., & Douglas, S. C. (2002). Toward an integrative theory of counterproductive workplace behavior: A causal reasoning perspective. *International Journal of Selection and Assessment, 10*, 36–49.

Martinko, M. J., & Moss, S. E. (2000). *Conflict in the leader-member dyad: An attributional interpretation.* Presented at the annual meeting of the Southern Management Association, Orlando, FL.

Martinko, M. J., Moss, S. E., & Douglas, S. C. (2003). *The effects of attribution styles on leader-member relations.* Presented at the annual meeting of the Academy of Management, Seattle, WA.

Martinko, M. J., & Zellars, K. (1998). Toward a theory of workplace violence: A cognitive appraisal perspective. In R. W. Griffin, A. O'Leary-Kelly, & J. M. Collins (Eds.), *Dysfunctional behavior in organizations: Violent and deviant behavior.* Stamford, CT: JAI Press.

Moberg, P. J. (2001). Linking conflict strategy to the five-factor model. Theoretical and empirical foundations. *International Journal of Conflict Management, 12.*

Nasby, W., Hayden, B., & DePaulo, B. M. (1979). Attributional bias among aggressive boys to interpret unambiguous social stimuli as displays of hostility. *Journal of Abnormal Psychology, 89*, 459–468.

Nicotera, A. M. (1995). *Conflict and organizations: Communicative processes.* Albany: State University of New York Press.

Pelted, L. H., Eisenhardt, K. M., & Xin, K. R. (1999). Exploring the black box: An analysis of work group diversity, conflict, and performance. *Administrative Science Quarterly, 44*, 1–28.

Perlow, R., & Latham, L. L. (1993). Relationship of client abuse with locus of control and gender. A longitudinal study in mental retardation facilities. *Journal of Applied Psychology, 78*, 831–834.

Pondy, L. R. (1967). Organizational conflict: Concepts and models. *Administrative Science Quarterly, 12*, 296–320.

Pruitt, D. G. (1983). Strategic choice in negotiation. *American Behavioral Scientist, 27*, 167–194.

Putnam, L. L. (1994). Beyond third party roles: Disputes and managerial intervention. *Employee Responsibilities and Rights Journal, 7*(1), 23–36.

Rahim, M. A. (1983). A measure of styles of handling interpersonal conflict. *Academy of Management Journal, 26*, 368–376.

Rahim, M. A. (1985). A strategy for managing conflict in complex organizations. *Human Relations, 38*, 81–89.

Rahim, M. A. (1997). Styles of managing organizational conflict: A critical review and synthesis of theory and research. In M. A. Rahim, R. T. Golembiewski, & L. E. Pate (Eds.), *Current topics in management* (Vol. 2, pp. 61–77). Greenwich, CT: JAI Press.

Rahim, M. A. (2000). *Managing conflict in organizations* (2nd ed.). Westport, CT: Quorum Books.

Rahim, M. A. (2001). *Managing conflict in organizations* (3rd ed.). Westport, CT: Quorum Books.

Rahim, M. A., & Bonoma, T. V. (1979). Managing organizational conflict: A model for diagnosis and intervention. *Psychological Reports, 44*, 1323–1344.

Rahim, M. A., & Magner, N. R. (1995). Confirmatory factor analysis of the styles of handling interpersonal conflict: First-order factor model and its invariance across groups. Journal of Applied Psychology, 1, 122–132.

Rahim M. A., Magner N. R., Shapiro D. L. (2000). Do justice perceptions influence styles of handling conflict with supervisors?: What justice perceptions, precisely. *International Journal of Conflict Management, 1*, 9–31.

Retzinger, S. M. (1991). *Violent emotions: Shame and rage marital quarrels.* Newbury Park, CA: Sage.

Russell, D. W. (1991). The measurement of attributional process: Trait and situational approaches. In S. L. Zelen (Ed.), *New model, new extensions of attribution theory* (pp. 99–108) New York: Springer-Verlag.

Sheppard, B. H. (1984). Third-party conflict intervention: A procedural framework. In B. Staw & L. L. Cummings (Eds.), *Research in organizational behavior* (Vol. 6, pp. 141–190). Greenwich, CT: JAI Press.

Snyder, M., & Ickes, W. (1985). Personality and social behavior. In G. Lindzey & E. Aronson (Eds.), *Handbook of social psychology* (3rd ed., pp. 883–947). New York: Random House.

Sternberg, R. J., & Dobson, D. M. (1987). Resolving interpersonal conflicts: An analysis of stylistic consistency. *Journal of Personality and Social Psychology, 52*, 794–812.

Sternberg, R. J., & Soriano, L. J. (1984). Styles of conflict resolution. *Journal of Personality and Social Psychology, 47*, 115–126.

Thomas, K. W. (1976). Conflict and conflict management. In M. D. Dunnette (Ed.), *Handbook of industrial and organizational psychology* (pp. 889–935). Chicago: Rand McNally.

Thomas, K. W. (1979). Conflict. In S. Kerr (Ed.), *Organizational behavior* (pp. 151–181). Columbus, OH: Grid.

Trevino, L. K., & Victor, B. (1992). Peer reporting of unethical behaviors: A social context perspective. *Academy of Management Journal, 35,* 38–64.

Tutzauer, F., & Roloff, M. E. (1988). Communication processes leading to integrative agreements: Three paths to joint benefits. *Communication Research, 15,* 360–380.

Utne, M. K., & Kidd, R. F. (1980). Attribution and equity. In G. Mikula (Ed.), *Justice and social interaction* (pp. 63–93). New York: Springer-Verlag.

Van de Vliert, E. (1997). *Complex interpersonal conflict behavior. Theoretical frontiers.* Hove, East Susex, UK: Psychology Press.

Van de Vliert, E., & Kabanoff, B. (1990). Toward theory-based measures of conflict management. *Academy of Management Journal, 33,* 199–209.

Vigil-King, D. C. (2000). *Team conflict, integrative conflict-management strategies, and team effectiveness: A field study.* Unpublished doctoral dissertation, University of Tennessee, Knoxville.

Wall, J. A., Jr., & Callister, R. R. (1995). Conflict and its management. *Journal of Management, 21,* 515–558.

Weiner, B. (1985). An attributional theory of achievement motivation and emotion. *Psychological Review, 92,* 548–573.

Weiner, B. (1995). *Judgments of responsibility.* New York: Guilford Press.

Wong, P. T. P., & Weiner, B. (1981). When people ask "why" questions and the heuristics of attributional search. *Journal of Personality and Social Psychology, 40,* 650–663.

CHAPTER 12

AN ATTRIBUTION–EMPATHY APPROACH TO CONFLICT AND NEGOTIATION IN MULTICULTURAL SETTINGS

Hector Betancourt
Loma Linda University

ABSTRACT

This chapter examines the role of attribution processes, emotions, and culture as antecedents to conflict resolution and negotiation. Specifically, White's (1985, 1987, 1991) realistic empathy approach to conflict and negotiation is analyzed from the perspective of an attribution–emotion approach to conflict. A model of prosocial behavior (Betancourt, 1990) and a related attribution–emotion model of conflict and violence (Betancourt, 1991; Betancourt & Blair, 1992) provide a conceptual frame for understanding the role of realistic empathy in conflict resolution and negotiation. In addition, data from studies on culture and attribution processes in conflict behavior (Zaw & Betancourt, 2002) are used to illustrate the importance of cultural and psychological factors in multicultural conflict environments.

Conflict is a natural aspect of human relations and it takes place at all levels of interaction, from interpersonal to institutional to international settings. From an economic perspective, the failure to resolve or prevent

Attribution Theory in the Organizational Sciences, pages 243–256

conflict may have important costs for society as well as for business and other institutions. From an organizational perspective, the direct costs associated with conflicts, such as those among employees, departments, different levels of administration, and labor organizations, are easy to identify. However, beyond such costly conflicts, corporations and institutions can also suffer the consequences of unresolved social or international conflicts. Moreover, in addition to suffering the consequences, corporations can often be the cause of political or social conflict at the local as well as international levels. Hence, understanding and effectively dealing with conflict is important not only to the individuals, groups, or countries directly involved but also to the local and international business community.

From a psychological perspective, although conflict and conflict negotiation may change from one setting to another (e.g., from interpersonal to international), some of the psychological processes and behaviors involved are relevant in any setting. Of course, since much psychological knowledge is based on laboratory research conducted at the individual or small-group level, one must be cautious in generalizing to large groups or international phenomena.

The aim of this chapter is to shed light on the role of psychological (attribution–emotion) processes in conflict behavior and variations associated with culture. First, Ralph White's (1985, 1987, 1991) realistic empathy approach to conflict resolution and negotiation is examined from the perspective of a model (Betancourt, 1990) integrating findings from attribution theory and the empathy approach to prosocial behavior. This attribution–empathy model and the research supporting it illustrate the nature of the relations among attribution processes and empathic feelings as determinants of behaviors relevant to effective conflict resolution and negotiation. Then, a second model is presented, in which anger is added to empathic emotions and attributions of intentionality are added to controllability of causal inferences as determinants of responses to conflict situations (Betancourt, 1991, 1997; Betancourt & Blair, 1992). This model, which builds on the previous one, provides a more comprehensive view of the attribution–emotion processes involved in conflict and negotiation. Finally, results from more recent studies (e.g., Zaw & Betancourt, 2002) are used to illustrate how culture influences conflict and negotiation not only directly, but also through its influence on attribution processes and related emotions.

THE REALISTIC EMPATHY APPROACH
TO CONFLICT AND NEGOTIATION

Realistic empathy (e.g., White, 1985, 1987, 1991) is defined as the under-standing of the situation of others as if one were looking at a conflict through their eyes. The focus is on the situation, not the other party as an individual or group, to the point that the causal inferences made for the actions of the others are seen as more situational than dispositional. When individuals see the conflict from the perspective of others, they can see all the options available to them, very much like in the case of a chess player.

According to the realistic empathy approach, empathic understanding elicits prosocial interactions and cooperation, which has a positive impact on effective negotiation and conflict resolution. Hence, in dealing with conflict it is important to maximize the ability of the parties in conflict to take the perspective of the other party, as a means of achieving empathic understanding. This process of perspective taking is conceived as a purely cognitive effort. In fact, although White recognizes that emotions are likely to play a role, he indicates that, as in the case of the chess player's type of empathy, emotions are not necessary to achieve an understanding of the other party's situation and options. While this approach has been found to be useful in dealing with international conflict (e.g., White, 1985, 1991), a conceptual frame integrating some of the psychological processes associ-ated with realistic empathy may enhance research and intervention in this area. The following sections, dealing with the role of attribution processes, empathic emotions, and culture, represent an effort in that direction.

PERSPECTIVE TAKING, ATTRIBUTIONS, EMPATHIC
EMOTIONS, AND PROSOCIAL BEHAVIOR

The association between empathy and prosocial behavior has long been documented (for reviews, see Davis, 1994; Eisenberg & Miller, 1987). Research in this area is diverse and includes phenomena such as the behav-ioral consequences of empathic perspective taking and the role of emo-tions (e.g., Batson, O'Quin, Fultz, Vanderplas, & Isen, 1983; Batson, Turk, Shaw, & Klein, 1995) to the effects of empathy on moral judgment and social justice (e.g., Davis, 2001; Hoffman, 1987, 1989, 2000; Murphy & Eisenberg, 2002).

According to Batson, empathic concern is characterized by the presence of what he defines as true empathic feelings (e.g., sympathy and compas-sion), as opposed to feelings of distress. These empathic feelings make prosocial behavior, in general, and altruism, in particular, possible (see Bat-son et al., 1983). According to this, higher levels of empathic emotions, such as sympathy and compassion, result in higher levels of prosocial behavior. In the case of conflict, this means that when individuals take the perspective of the others, they can experience more empathic feelings and thus may be more likely to cooperate in negotiating a positive and mutually beneficial resolution to the conflict.

From an attribution theory perspective (for a review, see Weiner, 1995), the causal inferences individuals make for the behavior or condition (e.g., need) of the other will influence whether or not they cooperate or responds in a prosocial way, such as helping. Specifically, in the case of a coworker who needs help, one is more likely to help if the need is attributed to uncontrollable versus controllable causes. As in the case of the empathy approach to prosocial behavior, the effect of the cognitive (attribution) process has been found to be mediated by interpersonal emotions.

Figure 12.1 represents a causal model integrating perspective taking, causal attributions concerning the needs of others, and empathic emotions as determinants of prosocial behavior. Specifically:

1. The paths from perspective to empathic emotions and from empathic emotions to prosocial behavior represent the view that what perspective taking does is to elicit empathic feelings, which are in turn the more direct determinants of prosocial behavior (e.g. Batson et al., 1995).

2. The paths from the need condition to the attribution process and from the attribution process to prosocial behavior indicate that aspects of the situation influence the attribution process concerning the causes of the need, which in turn influences prosocial behavior. The attribution process also influences feelings, which in turn influence behavior (for a review, see Weiner, 1995).

3. The path from perspective taking to the attribution process represents the proposition that perspective taking may influence the attribution process, in this case perception of controllability. This implies that taking the perspective of the other, which, according to White, is the key process involved in achieving realistic empathy, elicits empathic feelings and lowers the perceived controllability of the

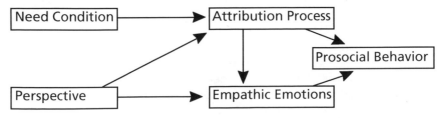

Figure 12.1.   An attribution–empathy model of prosocial behavior.

attributions one makes for the condition of the other, both of which increase the likelihood of prosocial responses.

4. Finally, the path from the attribution process to prosocial behavior represents the proposition that cognitive processes (e.g., attribution of controllability), may in some cases directly influence prosocial behavior, independent of an emotion-mediated effect. This is consistent with White's views that sometimes, as in the case of the chess player's kind of empathy, the cognitive processes may directly influence one's responses, independent of emotions.

The results from testing the causal model represented in Figure 12.1 (see Betancourt, 1990), provide a good fit for the experimental data, χ^2 (150, 3) = 2.37, p = .499. These results confirmed that adopting an empathic perspective influences empathic emotions. Higher levels of empathic feelings are in part due to the perception of lower controllability of attributions that results from adopting an empathic perspective. Thus, higher levels of prosocial behavior are a function of both higher levels of empathic feelings and lower levels of perceived controllability of the attributions concerning the conditions of the other party.

These results suggest that at least in part the effect of realistic empathy on conflict is mediated by the empathic feelings elicited by perspective taking. Moreover, since taking the perspective of the other also changes the attributions one makes for their condition, these results support White's views that realistic empathy involves a particular pattern of attributions, which he characterized as more situational than dispositional (see White, 1991). If this is so, other factors that influence empathic feelings and/or attribution processes may have similar effects on prosocial behavior, in general, and conflict, in particular.

FROM PROSOCIAL BEHAVIOR TO ANTISOCIAL (AGGRESSIVE) RESPONSES TO CONFLICT

The attribution–emotion model of conflict and violence presented in Figure 12.2 is based on the proposition that some of the same cognitive and emotional factors found to influence prosocial behavior may influence aggressive behavior during conflict. If the presence of empathic emotions increases prosocial behavior, such as cooperation and helping, while the absence of these emotions results in neglect, it is possible that high levels of empathic emotion may also inhibit aggressive responding. In addition, anger, the key determinant of aggression in the reformulated version of the frustration–aggression hypothesis (e.g., Berkowitz, 1983), has been found to be influenced by attribution processes similar to those affecting

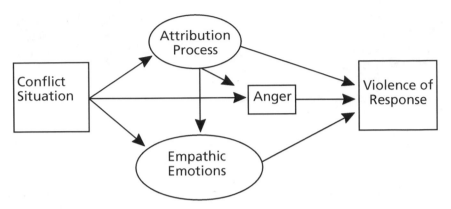

Figure 12.2.   Cognition (attribution)–emotion model of conflict and violence.

empathic emotions (e.g. Betancourt & Blair, 1992). In this case, attributing higher levels of controllability and intentionality to the negative actions of the other party results in higher levels of anger (for a review, see Weiner, 1995). Thus, while the model of Figure 12.1 is relevant to understanding prosocial behavior and a positive resolution of a conflict, the model in Figure 12.2 is relevant to understanding aggressive behavior, conflict escalation, and related negative outcomes. Specifically:

1. The paths from the frustrating conflict condition to anger and from anger to aggressive responding represent the traditional view of the frustration–aggression hypothesis, as reformulated by Berkowitz (1983).

2. The paths from the conflict condition to the attributions of controllability and intentionality, and from controllability and intentionality to anger, represent the view from attribution theory that anger is influenced by thinking processes concerning the causes of the conflict.

3. Consistent with the results for the prosocial behavior model (Figure 12.1), the path from the attribution process to empathic emotions represents the proposition that controllability of attributions influences empathic emotions. In addition, the path from the conflict condition to empathic emotions represents the proposition that, in addition to eliciting anger, aspects of the conflict situation directly influence empathic feelings, independent of the controllability and intentionality mediated effects.

4. The path from empathic emotions to aggressive responding represents the inhibiting effect of empathic feelings on aggressive responding. This is consistent with the view that in conflict environ-

ments empathy may not only have a positive effect on prosocial behaviors but also a negative effect on antisocial behavior.

5. Also consistent with research on prosocial behavior, the path from the attribution process to aggressive responding represents the view that cognition, in this case the perception of intentionality and controllability, may directly influence behavior, independent of the emotion mediated effects.

A test of this model was conducted using data from laboratory studies in which individuals were presented with a frustrating conflict situation (see Betancourt & Blair, 1992). The various experimental conditions manipulated the intentionality of the conflict behavior and the controllability of the causes to which that behavior was attributed. Figure 12.3 includes the results from this experimental test of the model, using Bentler's (1995) program for the analysis of structural equations (EQS). As observed in Figure 12.3, the model fits the data very well, NFI = .997, χ^2 (154, 9) = 11.48, p = .24.

Overall, the results show that the structure of relations among the cognitive processes and emotions involved during conflict is similar to the one observed in the attribution–emotion model of prosocial behavior presented in Figure 12.1. Specifically, when the instigating action is perceived as less intentional and attributed to less controllable causes, individuals experience less anger and more empathic emotions. While higher levels of anger result

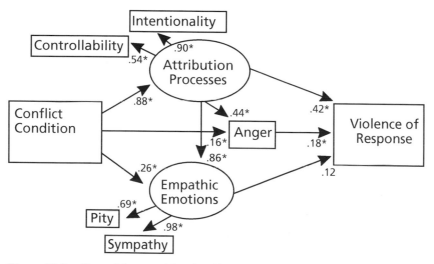

Figure 12.3.   Test of the cognition (attribution)–emotion model of conflict and violence. NFI = .977; χ^2 (154, 9) = 11.48, p = .2. From Betancourt & Blair (1992).

in a higher probability of an aggressive response, higher levels of empathic emotions result in a lower probability of aggressive responding.

Considering the purpose of this chapter, it is particularly interesting that the results highlight the limitations of traditional approaches that ignore the role of cognitive processes, such as attributional thinking and emotions other than anger. As demonstrated, cognitive processes not only influence empathic emotions but also anger and the probability of a positive resolution versus a violent outcome or escalation of a conflict.

Culture and the Cognition–Emotion Processes in Conflict Resolution

Changes in national demographics and globalization have created awareness of the need to better understand the role of culture in psychological functioning and behavior. Cultural diversity in the workplace, communities, and markets around the world significantly contribute to the complexities of communication and interaction among individuals, groups, institutions, and nations. A better understanding of how culture influences psychological processes and behavior at all levels appears to be necessary to promote cooperation and to more effectively manage negotiation and conflict resolution. Consistent with this view, a series of studies have been conducted recently (e.g., Betancourt & Zaw, 2003; Zaw & Betancourt, 2003) to examine aspects of culture, such as cultural value orientations, that may influence conflict behavior, such as preference for different styles of dealing with conflict. Results from that research are presented here to illustrate how cultural factors are likely to influence behaviors relevant to conflict resolution and negotiation, as well as the psychological processes likely to mediate this influence.

The models presented in Figures 12.4 and 12.5 are based on a set of theory-based relations among the collectivism/individualism value orientations, as examples of the kinds of cultural factors thought to influence conflict, attributions, and emotions as antecedents of conflict behavior. Specifically:

1. The paths from controllability to empathic emotion, from intentionality to anger, and from each of these two emotions to preference for a style of conflict resolution represent the proposition that the structure of relations among these variables is similar to that observed in the attribution–empathy model of prosocial behavior (Figure 12.1) and the attribution–emotion model of conflict and violence (Figure 12.3).

2. The direct paths from collectivism and individualism to conflict resolution style represent the results of previous research suggesting that

collectivists and individualists have preferences for different styles of conflict resolution, such as integrating or compromising versus dominating, respectively (Gabrielidis, Ibarra, Pearson, & Villareal, 1997; Itoi, Ohbuchi, & Fukuno, 1996; Pearson & Stephan, 1998).

3. The paths from collectivism and individualism to attributions concerning intentionality of the action and controllability of its cause represent the effects of culture on the attribution process and its mediating role in influencing conflict behavior.

Figures 12.4 and 12.5 include the results from testing this set of relations as antecedents of preference for different styles of conflict resolution, using Bentler's (1995) programs for the analysis of structural equations (EQS). Specifically, while the first model includes relations among collectivism, individualism, attribution processes, anger, and empathic emotions as predictors of preference for a dominating style, the second includes the same variables as antecedents of a compromising style. As observed in Figures 12.5 and 12.6, the results from testing these models fit the data for both the dominating and the compromising styles, respectively, with minor variations in some of the relations among the antecedent variables.

In general terms, the results concerning preference for a dominating style of conflict resolution (see Figure 12.4) show that individualism, but not collectivism, directly influences the preference for this style, independent of the cognition and emotion mediated effects. In addition, it shows

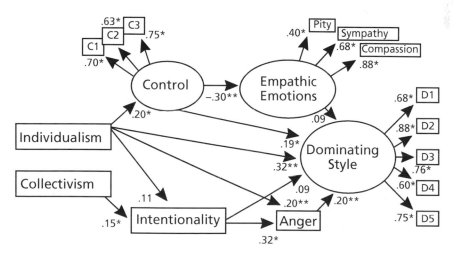

Figure 12.4.   Test of a model for culture and cognition (attribution)–emotion processes in conflict resolution: Dominating resolution style. From Zaw & Betancourt (2002). CFI = .967; χ^2 (83) = 105.61, p = .05.

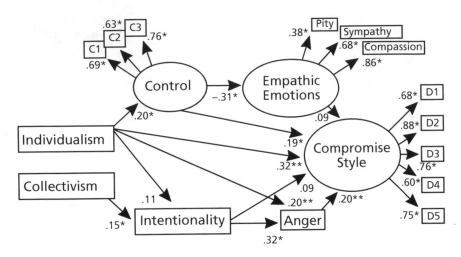

Figure 12.5.   Test of a model for culture and cognition (attribution)–emotion processes in conflict resolution: Compromising resolution style. From Zaw & Betancourt (2002). CFI = .942; $\chi 2$ (71) = 100.06, p = .013.

that individualism but not collectivism influences the perception of controllability, which in turn has a positive effect on preference for the dominating style. However, although collectivism does not appear to influence perception of controllability, it does have a positive effect on attribution of intentionality, which influences preference for the dominating style, both directly and through anger. In relation to the role of emotions, it is interesting that while anger significantly increases preference for a dominating style, empathic emotions do not seem to influence it significantly.

Concerning preference for a compromising style of conflict resolution, as observed in Figure 12.6, the model fits the data. However, there are some differences when compared to the model for the dominating style (Figure 12.5). Specifically, collectivism, which did not influence preference for a dominating style directly, does influence preference for a compromising style. Also, even though individualism shows an effect on the compromising style, this effect is negative, indicating that individuals who are higher on individualism demonstrate less preference for compromising than those who are low in individualism. In contrast, the higher individuals score on collectivism, the more likely they are to prefer a compromising style.

Finally, the test of these models suggests that preference for one or another style of conflict resolution is associated with different emotions. Specifically, according to these results, while anger plays a more important role influencing preference for a dominating style, empathic emotions, and not anger, influence the preference for compromising. Consistent with previous research, a higher level of empathic feeling, which is associated with prosocial behavior and nonviolent responding in conflict situations (see Figures 12.1 and 12.3), also results in preference for a compromising style.

CONCLUSIONS

In general, the attribution models of prosocial behavior (Figure 12.1) and violence in conflict situations (Figure 12.2) examined here provide a conceptual understanding of the processes involved in the realistic empathy approach to conflict resolution and negotiation (White, 1985, 1987, 1991). The research testing these models provides experimental evidence specifying the role of realistic empathy and the psychological processes involved in conflict behavior. For instance, the results from testing the attribution–empathy model of prosocial behavior (Betancourt, 1990) suggest that empathic perspective taking, the method proposed by White to induce realistic empathy, does in fact have a positive impact on prosocial behavior. Since these behaviors (e.g., cooperation and helping) are directly relevant to positive conflict resolution, the research supports the view that inducing empathy has a positive influence on conflict resolution and negotiation. In addition, the research demonstrates the role of the psychological processes involved in empathy-dependent conflict behavior.

Concerning the test of the attribution–emotion model of conflict and violence (Betancourt & Blair, 1992), results demonstrate that, in addition to empathic emotions, the attribution processes involved in conflict are likely to elicit other emotions, such as anger, which also influence responses to the conflict. Along with the studies of prosocial behavior, this research illustrates the kinds of psychological processes involved and how these affect conflict behavior. The conceptual understanding and experimental demonstration of the role of psychological processes involved in conflict behavior should allow for more effective interventions dealing with negotiation and conflict resolution.

In addition to the role of perspective taking in conflict behavior, the research examined the mediating role of attributional thinking and empathic feelings. These psychological processes, both of which appear to be natural components of empathic understanding, were recognized by White as being relevant. However, as these had not been systematically analyzed within the context of the realistic empathy approach, their role in conflict resolution and negotiation may have been underestimated.

Concerning the studies on the role of culture, such as the collectivistic and individualistic value orientations, these confirm the importance given by White to the need to understand others' and one's own cultural background. As observed in Figures 12.4 and 12.5, the data suggest that culture not only influences conflict behavior directly, but it also influences cognitive processes associated with empathic emotions and anger. Hence, consistent with research in other behavioral domains, the cognitive and emotional processes involved in prosocial behavior and violent responding appear to mediate, in part, the effects of culture on conflict behavior. This is particularly important given the increasing globalization of the economy, the demographic changes taking place in the United States, and the increasing cultural diversity observed in the workplace.

It is important to note that not all conflict behavior is affected by the same aspect of culture in the same way. For example, although individualism positively influences preference for one style of conflict resolution (e.g., dominating), this value orientation does not directly affect preference for other styles (e.g., compromising). This illustrates the importance for research on the role of culture in conflict resolution to identify and measure the specific aspects of culture that are most relevant to any given situation. Identifying and measuring such relevant cultural variables will allow to systematically analyze how these factors may relate to various behaviors in conflict and negotiation.

Of course, there are a number of other cultural and social psychological factors, from individual differences to situational, which are likely to influence conflict behavior. Hence, more research and more comprehensive theoretical models are necessary to effectively guide research and intervention in this area. This work represents an attempt to make a modest contribution in that direction. The psychological and cultural aspects included here are thought to be particularly relevant in conflicts where cultural and social identity factors can make a difference in the ability to empathize with the other party. This would be the case particularly in conflicts or disputes that take place in multicultural settings, from interpersonal and organizational to international.

REFERENCES

Batson, C., O'Quin, K., Fultz, J., Vanderplas, M., & Isen, A. M. (1983). Influence of self-reported distress and empathy on egoistic versus altruistic motivation to help. *Journal of Personality and Social Psychology, 45,* 706–718.

Batson, C., Turk, C., Shaw, L., & Klein, T. (1995). Information function of empathic emotions: Learning that we value the other's welfare. *Journal of Personality and Social Psychology, 68,* 300–313.

Berkowitz, L. (1983). The experience of anger as a parallel process in the display of impulsive, "angry" aggression. In R.Geen & E. Donnerstein (Eds.), *Aggression: Theoretical and empirical reviews: Vol. 1. Theoretical and methodological issues* (pp. 103–133). Orlando, FL.: Academic Press.

Bentler, P. (1995). *EQS structural equations program manual* [Computer software]. Encino, CA: Multivariate Software.

Betancourt, H. (1990). An attribution-empathy model of helping behavior: Behavioral intentions and judgments of help giving. *Personality and Social Psychology Bulletin, 16,* 573–591.

Betancourt, H. (1991). An attribution approach to intergroup and international conflict. In S. Graham & V. Folks (Eds.), *Attribution theory: Applications to achievement, mental health, and conflict.* Hillsdale, NJ: Erlbaum.

Betancourt, H. (1997). An attribution model of social conflict and violence: From psychological to intergroup phenomena. *Psykhe, 6,* 3–12.

Betancourt, H., & Blair, I. (1992). A cognition (attribution)-emotion model of violence in conflict situations. *Personality and Social Psychology Bulletin, 18,* 343–350.

Betancourt, H., & Zaw, G. (2003). Culture, attribution process, and conflict in multicultural educational settings. In F.Salili & R. Hoosain (Eds.), *Teaching, learning, and motivation in a multicultural context* (pp. 67–90). Greenwich, CT: Information Age.

Davis, M. H. (1994). *Empathy: A social psychological perspective.* Madison, WI: Brown and Benchmark.

Davis, M. H. (2001). Toward a comprehensive empathy-based theory of prosocial moral development. In A C. Bohart & D. J. Stipek (Eds.), *Constructive and destructive behavior: Implications for family, school, and society* (pp. 61–86). Washington, DC: American Psychological Association.

Eisenberg, H., & Miller, P. A. (1987). The relation of empathy to prosocial and related behaviors. *Psychological Bulletin, 101,* 91–119.

Gabrielidis, C., Stephan, W., Ybarra, O., Pearson, V. M., & Villareal, L. (1997). Preferred styles of conflict resolution: Mexico and United States. *Journal of Cross-Cultural Psychology, 28,* 661–677.

Hoffman, M. L. (1987). The contribution of empathy to justice and moral judgment. In N.Eisenberg & J. Strayer (Eds.), *Empathy and its development* (pp. 47–80). New York: Cambridge University Press.

Hoffman, M. L. (1989). Empathy and prosocial activism. In N. Eisenberg, J. Reykowski, & E. Staub (Eds.), *Social and moral values: Individual and societal perspectives* (pp. 65–85). New York: Erlbaum.

Hoffman, M. L. (2000). *Empathy and moral development: Implications for caring and justice.* New York: Cambridge University Press.

Itoi, R., Ohbuchi, K. I., & Fukuno, M. (1996). A cross-cultural study of preference of accounts: Relationship closeness, harm severity, and motives of account making. *Journal of Applied Social Psychology, 26,* 913–934.

Murphy, B., & Eisenberg, N. (2002). An integrative examination of peer conflict: Child reported goals, emotions, and behaviors. *Social Development, 11,* 534–557.

Pearson, V. M., & Stephan, W. (1998). Preferences for styles of negotiation: A comparison of Brazil and the U.S. *International Journal of Intercultural Relations, 22,* 67–83.

Weiner, B. (1995). *Judgments of responsibility.* New York: Springer-Verlag.

White, R. (1985). The psychological contributions to the prevention of nuclear war. *Applied Social Psychology Annual, 6,* 45–61.

White, R. (1987). The kinds of empathy needed in arms control negotiation. *American Journal of Social Psychiatry, 7,* 181–184.

White, R. (1991). Empathizing with Saddam Hussein. *Political Psychology, 12,* 291–184.

Zaw, G., & Betancourt, H. (2002). *Culture, attribution processes, and styles of conflict resolution.* Irvine, CA: Western Psychological Association.

CHAPTER 13

ANTECEDENTS TO DISSATISFACTION WITH AN INTERNATIONAL JOINT VENTURE PARTNER

The Role of Equity Theory and Attribution Theory

Daniel Laufer
University of Cincinnati

Byung Hee Lee
Hanyang University

ABSTRACT

International joint venture ("IJV") instability has been an important area in the past few years for both researchers and practitioners. This chapter introduces a new framework based on the satisfaction literature to analyze the causes of IJV instability. The framework proposes that dissatisfaction with a joint venture partner plays a major role in IJV instability. Two key determinants of dissatisfaction with a joint venture partner, equity and disconfirmation of expectations, are included in the framework and discussed. In addition, the

Attribution Theory in the Organizational Sciences, pages 257–273

chapter suggests ways to minimize the risk of dissatisfaction with a joint venture partner, thereby reducing the likelihood of IJV instability.

International joint venture ("IJV") instability has been an intriguing topic for both researchers as well as practitioners. IJV instability is defined in the literature in two ways. An outcome-oriented approach views IJV instability as a termination of IJVs or a change in the ownership structure of the IJV (Franko, 1971; Park & Ungson, 1997). A process-oriented approach, on the other hand, conceptualizes IJV instability as major reorganizations or contractual renegotiations (Yan & Zeng, 1999). Despite the different perspectives on IJV instability, many researchers as well as practitioners would agree that IJV instability can pose a serious threat for multinationals. Reported instability rates of IJVs in past studies have ranged from 25%–75% (Yan & Luo, 2001). Whereas entering a foreign country through an IJV has a number of important advantages including sharing the costs and risks of foreign entry as well as tapping into a partner's knowledge of the local environment including institutions, local consumer tastes, and business practices, problems cooperating with a partner from a different national culture in many instances dwarfs these benefits significantly and causes IJV instability.

Despite the importance of the topic to both researchers and practitioners, very few studies analyzing the connection between national culture and IJV instability have been conducted. The few studies that do exist in the literature have almost exclusively relied on Hofstede's (1991) typology of dimensions of national culture in attempting to explain the phenomenon. Unfortunately, this approach has not been successful in explaining IJV instability, as evidenced by the conflicting results obtained by different researchers (Barkema & Vermeulen, 1997; Li & Guisinger, 1991; Park & Ungson, 1997).

Consistent with a view that differences in the national culture of the partnering firms tend to be a threat to the stability of IJVs, Barkema and Vermeulen (1997) found a positive relationship between national cultural difference and IJV instability. In contrast, Park and Ungson (1997) reported that larger cultural distance was related to a lower IJV instability level.

This chapter proposes an alternative framework to Hofstede's typology in understanding the connection between national culture and IJV instability by incorporating the extensive literature on satisfaction (see Figure 13.1). This chapter introduces a key determinant of IJV instability, dissatisfaction with the JV partner, and identifies the major antecedents to this construct. It incorporates theories from social psychology impacting the dissatisfaction assessment, including equity theory and attribution theory, and explains how cultural differences impact this assessment. Finally, it

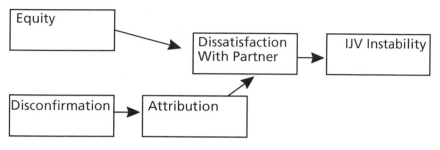

Figure 13.1.   Antecedents to IJV Instability [1]

discusses ways to prevent dissatisfaction from occurring, thereby limiting the threat of IJV instability.

IJV INSTABILITY

International joint ventures can be conceptualized as collaborative arrangements in which a separate legal entity is created by at least two parent firms in different countries that share ownership interests under conditions and provisions that are specified in a contract (Luo, 2002).

Although a joint venture is destined to be terminated at a prespecified temporal point lapsed from its inception, it is highly likely to experience unexpected or intentional changes in the agreed terms. These changes in contractual terms often render partnering firms reorganizing costs, alter the equitable relationship between partners, and thus result in growing IJV instability (Gray & Yan, 1997).

Yan and Zeng (1999) documented previous literature on IJV instability. Previous studies with a static and outcome-oriented perspective conceptualized IJV instability as termination or changes in ownership structure. This perspective was originated in an early study of Franko (1971) that examined ownership changes in IJVs formed by U.S. manufacturing firms. Following Franko's instability approach, other studies focused on unexpected IJV termination or dissolution as an indicator of instability (e.g., Barkema & Vermeulen, 1997; Park & Ungson, 1997).

On the other hand, research with a dynamic and process-oriented perspective viewed instability as reorganizations or contractual renegotiations. This stream of literature emphasized the structural and operational aspects of IJVs (e.g., Killing, 1983). A major contribution of these studies is the shift in focus from analyzing the final status of IJVs to examining factors conducive to IJV instability.

CULTURE AS A FACTOR CONTRIBUTING
TO IJV INSTABILITY

Since IJVs involve at least two parent firms located in different countries, there may exist joint venture management costs resulting from cultural differences of the partnering firms.

Cultural differences indicate the extent to which work-related assumptions between the firms converge (Hofstede, 1991). Similar cultural values possessed by partner firms can reduce coordination costs, misinterpretation of strategic issues, and miscommunications. Similarly, Lane and Beamish (1990) posit that different national cultures influence behavior and management systems of IJVs that may lead to unresolved conflicts. For instance, mechanisms enhancing cooperation take various forms between individualist and collectivist cultures due to the differences in their instrumental and expressive motives (Chen, Chen, & Meindl, 1998). Hence, IJVs involving firms with similar cultural values tend to be more stable than IJVs formed by firms located in culturally distant countries. However, empirical findings on cultural distance and IJV instability are mixed. Although Barkema and Vermuelen (1997) supported this positive relationship between cultural distance and IJV instability, Park and Ungson (1997) found an opposite relationship and Fey and Beamish (2001) found no significant association. These conflicting findings suggest that a new approach to IJV instability is warranted. In fact, a number of researchers have pointed to the need for further research in this area (Hennart & Zeng, 2002).

In the following section we present an alternative framework to understanding the IJV instability phenomenon. This approach incorporates the extensive literature on satisfaction and focuses on areas of friction between IJV partners that can potentially generate instability. In Shenkar's (2001) critique of the cultural distance construct, he suggests that the focus should be on areas of friction and not necessarily cultural differences. Not every cultural difference generates friction between partners and our proposed framework incorporates this approach.

MODEL OF SATISFACTION

Consumer satisfaction is an area of great interest in the field of marketing (Oliver, 1997; Yi, 1990). Tse and Wilton (1988, p. 204) define consumer satisfaction as "the consumer's response to the evaluation of the perceived discrepancy between prior expectations (or some norm of performance) and the actual performance of the product as perceived after its consumption." The importance of the consumer satisfaction construct has been demonstrated through numerous studies of the negative consequences of consumer dissatisfaction. These studies have linked consumer dissatisfaction with complaining behavior, future purchase intentions, and negative word of mouth activity (Folkes, 1984, 1988; Oliver, 1997).

The literature on consumer satisfaction in marketing provides a rich foundation from which propositions can be developed regarding IJVs as well. Recently, consumers have become increasingly involved in joint production together with firms (Bendapudi & Leone, 2003). This type of customer participation in co-production is conceptually similar to a joint venture between two firms and understanding how dissatisfaction is formed in one context may assist us in understanding the process in the other.

As previously mentioned, dissatisfaction plays a key role in consumers' behavior in a marketing context. Consumers form relationships with their brands through consumption, and dissatisfaction adversely impacts these relationships (Fournier, 1998). In an IJV, dissatisfaction with a partner can also adversely impact the relationship between the joint venture partners. In fact, Shamdasani and Sheth (1995) examined satisfaction in marketing alliances between partners from the United States and found that the correlation between alliance satisfaction and continuity with the joint venture was high (r = 0.86).

Therefore, Drawing on Tse and Wilton's (1988) definition from the marketing literature, we define satisfaction with a joint venture partner as "a partner's response to the evaluation of the perceived discrepancy between prior expectations (or some norm of performance) and the actual performance of the partner as perceived after the formation of a joint venture." In the IJV context, we would also expect that:

Proposition 1: *Dissatisfaction with the joint venture partner increases the likelihood of IJV instability.*

As suggested by the definition of satisfaction, satisfaction models entail a comparison process. In a recent meta-analysis of studies of consumer satisfaction, Szymanski and Henard (2001) found that the two antecedents with the greatest impact on the satisfaction construct were equity and the disconfirmation of expectations. In the following sections we describe these antecedents in detail and explain how they impact the dissatisfaction with IJV partner construct.

EQUITY

Equity is an evaluation of fairness that people make in the context of a relationship and is based on equity theory. Szymanski and Henard (2001), in their meta-analysis of customer satisfaction, found that equity was the antecedent with the greatest impact on customer satisfaction, with a mean correlation of .50 between the two constructs. Equity theory refers to a need mechanism that is activated when the ratio of resources received (rewards, etc.) to "investments" made is not equal to the ratio for others in the exchange relationship (Adams, 1963). Inputs are defined as "contributions to the exchange which are seen by the participant as entitling him to rewards" and outcomes are defined as "the positive and negative consequences that a participant has incurred as a consequence of his relationship with another (Walster, Walster, & Berscheid, 1978, p. 152). The activated need for equity stimulates behavior that is aimed at creating a more equitable distribution, that is, one in which investment/rewards ratios are equal. Adams (1963) also suggests that the state of perceived inequity creates tension, which a participant wishes to reduce. Methods of reducing perceived inequity might include changing one's inputs into the relationship, changing one's perception of one's outcomes from the relationship, or leaving the relationship (Walster et al., 1978). A number of researchers suggest, however, that instead of an intraindividual need, equity can be viewed as a shared cultural value, a norm that prescribes the way resources ought to be distributed, and we may find variation across cultural samples in the following areas (Pepitone & Triandis, 1987):

1. The conditions that activate, stop, and govern the critical magnitudes of equity and inequity—for example, groups may differ in the degree of disparity tolerated before distribution behavior is affected.

2. The ways people try to make the distributions equitable.

3. The criteria that count as investments and the measurements of rewards.

The importance of equity in arriving at satisfaction evaluations is by no means limited to the realm of consumer behavior. Perceptions of fairness play a big role in joint ventures as well. Partners evaluate each other in terms of carrying their fair share of the work. Perceptions of inequity can generate dissatisfaction with the JV partner, which in turn can create IJV instability.

The findings, however, regarding cross-cultural differences in equity theory suggest that the influence of the equity construct depends on the cultural background of the IJV partners. In high-power distance societies, individuals are more likely to accept hierarchical relations and inequity. Therefore, JV partners from these countries may be more likely to tolerate situations where their share of the work is greater compared with the other JV partners. There also may be cross-cultural differences in evaluating the efforts involved in an IJV. Different cultures may value inputs/outcomes differently. Functional equivalence, which involves determining whether the concepts, objects, or behaviors have the same role or function in all countries studied, can be difficult to find. A good example of a study that examines functional equivalence can be found in the marketing literature. Green and Alden (1988) found that gift giving in Japan and the United States is functionally different. Gift giving in Japan was found to serve an important affiliation function and was a more common occurrence compared with the United States where gift giving was less crucial to reinforcing an individualistic self-concept. The lack of functional equivalence can impact the perception of equity because functionally different inputs and outputs are valued differently in different countries, and thereby impact the evaluation of the exchange cross-culturally. In the case of IJVs, contributions made by the partners may be interpreted differently by the different sides due to dissimilarity of the partners' functional equivalence as well. A good example of this is the work involved in complying with government regulations in different countries. Whereas obtaining government approval to operate a business in the United States may be a relatively easy task, a similar task in China could be much more complex and time consuming. If the American partner compares the Chinese partners' efforts to his or her experience in the United States, an incorrect assessment may be made regarding the amount of work the Chinese partner is investing in the IJV. This in turn could generate a perception of inequity in the relationship, thereby increasing the likelihood of dissatisfaction with the partner and IJV instability.

Proposition 2: *Perceptions of inequity increase dissatisfaction with the joint venture partner.*

DISCONFIRMATION OF EXPECTATIONS AND ATTRIBUTIONS

The second antecedent with the greatest impact on customer satisfaction in Szymanski and Henard's (2001) meta-analysis of customer satisfaction is disconfirmation of expectations, with a mean correlation of .46 between the two constructs. The concept underlying the disconfirmation of expectations paradigm is that people reach satisfaction decisions by comparing performance with prior expectations. If performance fails to meet expectations, disconfirmation occurs. It is worth noting that disconfirmation by itself does not necessarily generate dissatisfaction. When disconfirmation occurs people seek reasons for the surprising outcome. These reasons or causal attributions mediate satisfaction assessments.

Research in the area of attribution theory suggests that cross-cultural differences exist in the area of attributions. Heider (1958) referred to two types of explanations that are given to explain the causes of events by people: (1) external attribution where the individual attributes the causes to environmental factors or (2) internal attribution where the causes are attributed to dispositional factors. In a joint venture context, this represents blaming the partner for the failure to achieve a certain goal. Heider found that people tend to overestimate a person's personal liability for his/her behavior and to underestimate the social and economic pressures that may contribute to it. This tendency has been defined as the fundamental attribution error (Ross, 1977). It is worth noting that these findings were based on studies conducted in individualistic societies primarily in the United States.

In contrast to the findings from the early studies on the fundamental attribution error, a number of recent studies outside of the United States have shown that the predominant tendency of observers to attribute personal or internal causes of an actor's behavior does not replicate. When describing themselves or others, studies have shown that Asians make more contextual references and fewer dispositional references than Europeans or Americans, suggesting a more contextualized theory of behavior (Choi, Nisbett, & Norenzayan, 1999). For example, Miller (1984) found that Indian middle-class adults primarily attribute the causes of deviant behaviors to external features of the social environment, the reverse pattern of that shown by a comparable sample of U.S. adults. In addition, in a recent review of studies comparing North American and East Asian perceivers, researchers concluded that the sharpest differences in attributions for the cause of an individual's behavior lie in the weight accorded to the contexts of constraints and pressures imposed by social groups (Choi et al., 1999). Choi and colleagues (1999) suggest that different thinking styles between Asian and Western cultures may explain some of the differences in attribution. Westerners use analytic thinking, paying attention primarily to the object, categorizing it on the basis of its attributes, and attributing causality to the object based on rules about its category memberships (Lloyd, 1990). In contrast, Asians perceive and reason holistically, attending to the field in which objects are embedded and attributing causality to interactions between the object and field (Choi et al., 1999). Choi and colleagues propose three plausible models to explain the differences in attribution styles between Asian versus Western cultures: (1) Asians may follow a sequence of situational inference followed by dispositional correction, whereas Americans follow a sequence of dispositional inference followed by situational corrections; (2) Asians may make more situational corrections than Americans, with little difference in dispositional inferences; and (3) the initial dispositional inference might be weaker for Asians than for

Americans. Researchers, however, have yet to determine which of the three models described above best explains the differences in attribution styles between individuals in Asian and Western cultures. An interesting study by Chiu, Morris, Hong, and Menon (2000) suggests that when the need for closure increases (manipulated by placing time constraints on subjects in determining attributions), the more likely subjects will rely on chronically accessible knowledge structures in a culture. The authors found in their study that when the subjects were under pressure, North American participants increased attributions to personal factors; however, the Chinese did not. These findings supported previous studies examining cross-cultural differences in attributions.

Another possible reason for cross-cultural differences in attribution styles is differences in levels of locus of control. Gilbert (1995), for example, suggests that dispositional attributions provide people with a sense of control whereas attributing a cause to the situation implies that the individual does not exert control over his or her situation. This of course assumes a culture with high levels of locus of control such as the United States (Cote & Tansuhaj, 1989). In countries with low levels of locus of control we would not expect the fundamental attribution error to occur as often, since individuals do not expect to have much influence over the situation.

Cross-cultural differences in the fundamental attribution error have important implications for IJV instability. If partners from individualistic societies such as the United States have a predisposition to blame the other partners for failures relating to the joint venture, this may generate dissatisfaction with those partners' performance and thereby increase IJV instability. Cross-cultural differences in the fundamental attribution error may partially explain the higher instability rates of Japanese IJVs in the United States (Hennart & Zeng, 1997) compared with the much lower rates of Japanese IJVs in Asia (Makino, 1995). Perhaps both the Japanese and other Asian partners are more likely to attribute failures relating to the IJV to situational factors such as economic conditions, thereby reducing the levels of dissatisfaction with the performance of the IJV partners.

Proposition 3: *The likelihood of the occurrence of dissatisfaction with an international joint venture partner increases when one or more of the partners are from countries where the fundamental attribution error is more likely to occur.*

COUNTRY OF ORIGIN STEREOTYPES AND ATTRIBUTIONS

In addition to cross-cultural differences in the fundamental attribution error, prior beliefs manifested through stereotypes may also impact attributions. A person's preexisting hypotheses and suppositions can impact attributions when disconfirmation occurs (Folkes, 1988). One type of prior belief is a stereotype. Stereotypes are widespread beliefs about social groups and many cognitive psychologists believe that stereotyping occurs as a result of biases in cognition, especially in the operation of perceptions and memory. In fact, Bodenhausen and Lichtenstein (1987) define stereotyping as a "simplification strategy employed by the social perceiver to facilitate his or her interactions with a complex environment."

Individuals learn stereotypes through the socialization process. The internalization of stereotypes occurs through primary influences such as family and secondary influences such as education and exposure to media (Janda & Rao 1997). Evoking a stereotype involves categorizing an individual into a group and then associating the characteristics of the group to the individual. Stereotypes discourage recognition of differences among members of targeted groups and therefore can have an adverse impact on social judgments, especially in cases of heterogeneous groups. It is worth noting that the activation of a stereotype is not necessarily a conscious activity and Devine (1989) found that common stereotypes are activated automatically when members of the stereotyped group are encountered.

Much of the evidence regarding the connection between attributions and stereotypes is derived from studies analyzing the role prejudicial views play within the judicial process. A great deal of research in this area has demonstrated that many social groups are perceived negatively and these prejudicial views may play a direct role within the judicial system. Preexisting beliefs about the defendant's social group may affect judgments of culpability and predictions of future criminal behavior, particularly if the crime is stereotypically linked to the defendant's social group. For example, it is well documented that Hispanics in the United States are often perceived as aggressive (Jones, 1991; Marin, 1984) and this stereotype may lead to discriminatory judgments in judicial proceedings that involve aggressive crimes such as criminal assault compared with Anglo-Americans (Bodenhausen, 1988; Bodenhausen & Lichtenstein, 1987).

In the context of an IJV, the country of origin (COO) of the partners is likely to act as a stereotype. Samiee (1994) defines a country stereotyping effect as "any influence or bias resulting from COO." The country that a partner originates from serves as an extrinsic cue to the other partners regarding variables that are unknown such as blame. As previously mentioned, the IJV partners perceive and adapt to opportunities and threats in their environment and can interpret things differently. For example, missing a sales forecast can be attributed to the poor performance of the partner or situational factors such as an economic slowdown. If the COO stereotype of the partner is of a negative nature, the other partner is more likely to attribute missing the sales forecast to the poor performance of the partner, which would be in congruence with the nature of the negative

COO. This expectation is supported by the extensive literature on COO in the field of international business, which suggests that COO acts as an informational stimulus about a product that is used by consumers to infer beliefs regarding product attributes such as quality, value, and perceptions of risk (Bilkey & Nes, 1982; Liefeld, 1993; Verlegh & Steenkamp, 1999). In addition to impacting product evaluations, a recent study found that the country of origin cue impacts the assessment of blame by consumers in a product-harm crisis (Laufer, 2002).

Proposition 4:   *The likelihood of the occurrence of dissatisfaction with a joint venture partner increases when one of the partners is from a country with a negative COO stereotype.*

MANAGERIAL IMPLICATIONS

In a recent review of the IJV instability literature (Yan & Zeng, 1999), the authors found that a major weakness in the work conducted in this area is the lack of practical relevance for managers. For example, in the studies incorporating Hofstede's (1991) measures, what should a company do if one dimension appears to impact IJV instability whereas another does not? Does this finding suggest that a company should not seek a partner from a country with an undesirable score on that particular cultural dimension? Are there ways to counterbalance the negative impact of the cultural dimension through training? These issues have not been addressed in the literature.

The dissatisfaction with partner framework not only provides us with a better understanding of how IJV instability occurs, but it also enables us to develop ways to minimize the occurrence of dissatisfaction, thereby decreasing the likelihood of IJV instability. The following are actions companies can take to reduce dissatisfaction with IJV partners.

Perceptions of Inequity

As previously described, certain cultures place more importance on an equitable division of effort between joint venture partners than others. In these instances the partners should emphasize the important role the other partner is playing in order to minimize the chance that perceptions of inequity will arise. The likelihood of IJV instability is also greatly enhanced if the roles that the partners are playing are not very well understood by the various sides. As previously mentioned, the concept of functional equivalence is difficult to find across cultures, so the time and effort invested in performing various tasks may be evaluated differently in different cultures. This highlights the importance of gaining a better understanding of the complexity of the tasks being performed by the various partners and communicating this to the partners' employees.

Disconfirmation of Expectations and Attributions

When performance does not meet expectations, dissatisfaction may occur. It is important that all partners fully understand the expectations regarding each other's role in the joint venture. Even though this may seem obvious, when partners come from different countries, the chances for misunderstandings are greatly increased.

As previously mentioned, disconfirmation does not in itself create dissatisfaction. Attributions regarding the perceived causes of the disconfirmation determine whether dissatisfaction with the IJV partner will occur. Recent research on attribution theory suggests that some cultures are more prone to making situational attributions than others (Choi et al., 1999). The implication of this is that events relating to the joint venture may be interpreted very differently by the various partners. Partners from individualistic cultures may be quicker to blame the other partner for not meeting a goal whereas an Asian partner may be more likely to attribute this to situational factors. It is very important for the IJV partners to recognize the existence of these cross-cultural differences in attributions. A partner from the United States may mistakenly believe that the Asian partner is trying to avoid taking responsibility for not meeting its goals, thereby creating an even worse situation by adversely impacting trust between the partners. Realizing that Asian cultures place a greater emphasis on situational factors would help the American side understand that attributions are influenced by culture, and the partner's situational attribution does not suggest a lack of willingness by the Asian partner to accept responsibility for failures. The fact that the American partner is more prone to attribute dispositional factors to failures is also important for Asian partners to realize. If Asian partners do not take this into account, they may incorrectly assume that disagreements over the causes of events relate to the lack of trust between the partners and not cultural differences in attributions. As with cross-cultural differences in the importance of equity, the partners should invest time and effort in developing training programs designed to educate employees of the cross-cultural differences in attributions in order to minimize potential friction between the partners.

Country of Origin Stereotypes and Attributions

As previously mentioned, prior beliefs can also impact attributions. In the context of an IJV, prior beliefs may be manifested through the COO stereotype. Research in the field of social psychology suggests that changing the nature of a negative stereotype is very difficult to accomplish, therefore it is better for a member of a stereotyped group to try to convince others that he or she is an exception to the rule (Hewstone, 1994). Perhaps in the context of an IJV, a JV partner can emphasize to its employees that the other JV partner from a negative COO is a positive exception to the companies typically found in that country. For example, the JV partner can emphasize to its employees that the other JV partner operates world-class facilities or that a number of nonlocal employees from a positive COO, for example, play key roles in the company. This may help convince employees in the IJVs that the partner from the negative COO is an exception to the perceived performance of a typical company from that country. Even if the employees do not perceive the company to be an exception to the negative COO stereotype, a co-branding effect may be generated by emphasizing the connection with a positive COO. Simonin and Ruth (1998) define co-branding as the "short or long term association or combination of two or more individual brands, products and/or distinctive proprietary assets." Brand image is defined as perceptions of the brand that reflect consumer associations of the brand in memory (Keller, 1993). When two brands are presented together, both evaluations are likely to be evoked by the consumer (Brioniarczk & Alba, 1994). In the previous example, both perceptions regarding the negative COO and the positive COO are evoked because the company is from a negative COO and the technology/employees are from a positive COO. This reduces the negative impact of only evoking the negative COO stereotype, which occurs when the information pertaining to the technology/employees from a positive COO is not emphasized by the partner.

CONCLUSION

The consumer satisfaction literature is a valuable source from which to draw on in attempting to shed light on the factors in IJV instability. Cross-cultural differences in attribution theory and equity theory provide clues to how dissatisfaction occurs between IJV partners and also provides guidance to companies regarding actions to take to minimize friction. Prior beliefs manifested through stereotypes can also adversely impact joint venture stability and the extensive literature on the COO effect suggests that the origin of a partner can impact assessments of dissatisfaction. The stereotyping literature from social psychology provides guidance with regard to the ways to counter a negative COO effect.

The dissatisfaction with an IJV partner framework enriches our understanding of IJV instability and explains the phenomenon more richly than Hofstede's framework. As previously mentioned, the concept of cultural distance is perceived by many as too simplistic and not effective in capturing the complexity of the IJV instability phenomenon. The dissatisfaction with an IJV partner framework has the potential to further our understanding of this extremely important phenomenon.

NOTES

1.   Our model depicts equity and attributions as independently impacting the dissatisfaction construct. It is worth noting that perceptions of equity and attributions generated from the disconfirmation of expectations may also impact each other. For example, if the joint venture partner attributes not meeting a goal (such as a sales forecast) to the inaction of the other joint venture partner, this may also impact perceptions of equity (Is the joint venture partner doing its fair share of the work?) Future research should also examine the interaction between these two variables and the resulting impact on the dissatisfaction with IJV partner construct.

REFERENCES

Adams, J. (1963). Toward an understanding of inequity. *Journal of Abnormal and Social Psychology, 67*, 422–436.

Barakema, H., & Vereulen, F. (1997). What differences in the cultural backgrounds of partners are detrimental for international joint ventures? *Journal of International Business Studies, 28*(4), 845–864.

Bendapudi, N., & Leone, R. (2003). Psychological implications of customer participation in co-production. *Journal of Marketing, 67*(1), 14–28.

Bilkey, W., & Nes, E. (1982). Country-of-origin effects on product evaluations. *Journal of International Business Studies, 13*(1), 89–99.

Bodenhausen, G. (1988). Stereotypic biases in social decision making and memory: Testing process models of stereotype use. *Journal of Personality and Social Psychology, 55*, 726–737.

Bodenhausen, G., & Lichtenstein, M. (1987). Social stereotypes and information processing strategies: The impact of task complexity. *Journal of Personality and Social Psychology, 52*, 871–880.

Broniarczyk, S., & Alba, J. (1994). The importance of brand in brand extension. *Journal of Marketing Research, 31*, 214–228.

Chen, C., Chen, X. P., & Meindl, J. (1998). How can cooperation be fostered? The cultural effects of individualsim-collectivism. *Academy of Management Review, 23*(2), 285–304.

Chiu, C., Morris, M., Hong, Y., & Menon, T. (2000). Motivated cultural cognition: The impact of implicit cultural theories on dispositional attribution varies as a function of need for closure. *Journal of Personality and Social Psychology, 7*(2), 247–259.

Choi, I., Nisbett, R., & Norenzayan, A. (1999). Causal attribution across cultures: Variation and universality. *Psychological Bulletin, 125,* 47–63.

Cote, J., & Tansuhaj, P. (1989). Culture bound assumptions in behavior intension models. *Advances in Consumer Research, 16,* 105–109.

Devine, P. (1989). Stereotypes and prejudice: Their automatic and controlled components. *Journal of Personality and Social Psychology, 56,* 5–18.

Fey, C., & Beamish, P. (2001). Organizational climate similarity and performance: International joint ventures in Russia. *Organization Studies, 22,* 853–882.

Folkes, V. (1984). Consumer reactions to product failure: An attributional approach. *Journal of Consumer Research, 10,* 398–409.

Folkes, V. (1988). Recent attribution research in consumer behavior: A review and new directions. *Journal of Consumer Research, 14,* 548–565.

Fournier, S. (1998). Consumers and their brands: Developing relationship theory in consumer research. *Journal of Consumer Research, 24,* 343–373.

Franko, L. (1971). *Joint venture survival in multinational corporations.* New York: Praeger.

Gilbert, D. (1995). Attribution and interpersonal perception. In A. Tesser (Ed.), *Advanced social psychology* (pp. 102–242). New York: McGraw-Hill.

Gray, B., & Yan, A. (1997). Formation and evolution of international joint ventures. In P. Beamish & J. P. Killing (Eds.), *Cooperative strategies: Asian Pacific perspectives* (pp. 57–88). San Francisco: New Lexington Press.

Green, R., & Alden, D. (1988). Functional equivalence in cross-cultural consumer behavior: Gift giving in Japan and the United States. *Psychology and Marketing, 5*(2), 156–168.

Heider, F. (1958). *The psychology of interpersonal relations.* New York: Wiley

Hennart, J., & Zeng, M. (1997). *Is cross-cultural conflict driving international joint venture instability? A comparative study of Japanese-Japanese and Japanese-American IJVs in the United States.* Paper presented at the annual meeting of the Academy of Management, Boston.

Hennart, J. F., & Zeng, M. (2002). Cross-cultural differences and joint venture longevity. *Journal of International Business Studies, 33*(4), 699–716.

Hewstone, M. (1994). Revision and change of stereotypic beliefs: In search of the elusive subtyping model. In W. Stroebe & M. Hewstone (Eds.), *European review of social psychology* (Vol 5, pp. 69–109). Chichester, UK: Wiley.

Hofstede, G. (1991). *Cultures and organizations: Software of the mind.* Berkshire, UK: McGraw-Hill.

Janda, J., & Rao, C. (1997). The effect of country-of-origin related stereotypes and personal beliefs on product evaluation. *Psychology and Marketing, 14*(7), 689–702.

Jones, M. (1991). Stereotyping Hispanics and whites: Perceived differences in social roles as a determinant of ethnic stereotypes. *Journal of Social Psychology, 131,* 469–476.

Keller, K. (1993). Conceptualizing, measuring, and managing customer-based brand equity. *Journal of Marketing, 57,* 1–22.

Killing, J. P. (1983). *Strategies for joint venture success.* New York: Praeger.

Kotler, P. (1989). *Principles of Marketing.* New York: Prentice Hall.

Lane, H., & Beamish, P. (1990). Cross-cultural cooperative behavior in joint ventures in LDCs. *Management International Review, 30*[Special Issue], 87–102.

Laufer, D. (2002). Product crises and consumers' assessment of blame: Is there an impact of country of origin? Unpublished doctoral dissertation, University of Texas at Austin.

Li, J., & Guisinger, S. (1991). Comparative business failures of foreign-controlled firms in the United States. *Journal of International Business, 22*(2), 209–224.

Liefeld, J. (1993). Experiments on country-of-origin effects: Review and meta-analysis of effect size. In N. Papadopoulos & L. Heslop (Eds.), *Product-country images: Impact and role in international marketing* (pp. 117–156). New York: International Business Press.

Lloyd G. (1990). *Demystifying mentalities.* New York: Cambridge University Press.

Luo, Y. 2002. Contract, cooperation, and performance in international joint ventures. *Strategic Management Journal, 23*(10), 903–919.

Makino, S. (1995). *Joint venture ownership structure and performance: Japanese joint ventures in Asia.* Unpublished doctoral dissertation, University of Western Ontario, London, Ontario, Canada.

Marin, G. (1984). Stereotyping Hispanics: The differential effect of research method, label, and degree of contact. *International Journal of Intercultural Relations, 8,* 17–27.

Miller, G. (1984). Culture and the development of everyday social explanation. *Journal of Personality and Social Psychology, 46,* 961–978.

Morgan, R., & Hunt S. (1994). The commitment–trust theory of relationship marketing. *Journal of Marketing, 58,* 20–38.

Oliver, R. (1997). *Satisfaction: A behavioral perspective on the consumer.* New York: McGraw-Hill.

Park, S., & Ungson, G. (1997). The effect of national culture, organizational complementarity, and economic motivation on joint venture dissolution. *Academy of Management Journal, 40*(2), 279–308.

Pepitone, A., & Triandis, H. (1987). On the universality of social psychological theories. *Journal of Cross-Cultural Psychology, 18*(4), 471–498.

Ross, L. (1977). The intuitive scientist and his shortcomings. In L. Berkowitz *Advances in experimental social psychology* (pp. 67–84). New York: Academic Press.

Samiee, S. (1994). Customer evaluation of products in a global market. *Journal of International Business, 23*(3), 579–603.

Shamdasani, P., & Sheth, J. (1995). An experimental approach to investigating satisfaction and continuity in marketing alliances. *European Journal of Marketing, 29*(4), 6–23.

Shenkar, O. (2001). Cultural distance revisited: Toward a more rigorous conceptualization and measurement of cultural differences. *Journal of International Business, 32*(3), 519–535.

Simonin, B., & Ruth, J. (1998). Is a company known by the company it keeps? Assessing the spillover effects of brand alliances on consumer brand attitudes. *Journal of Marketing Research, 35,* 30–42.

Szymanski, B., & Henard, D. (2001). Customer satisfaction: A meta-analysis of the empirical evidence. *Journal of the Academy of Marketing Science, 29*(1), 16–35.

Tse, D., & Wilton, P. (1988). Models of consumer satisfaction: An extension. *Journal of Marketing Research, 25*(3), 204–212.

Verlegh, P., & Steenkamp, J. (1999). A review and meta-analysis of country-of-origin research. *Journal of Economic Psychology, 20,* 521–546.

Walster, E., Walster, G., & Berscheid, E. (1978). *Equity: Theory and research.* New York: Springer-Verlag.

Yan, A., & Luo, Y. (2001). *International joint ventures: Theory and practice.* Armonk, NY: Sharpe.

Yan, A., & Zeng, M. (1999). International joint venture instability: A critique of previous research, a reconceptualization, and directions for future research. *Journal of International Business Studies, 30*(2), 397–414.

Yi, Y. (1990). A critical review of consumer satisfaction. In V. Zeithaml (Ed.), *Review of marketing* (pp. 35–78). Chicago: American Marketing Association.

## CHAPTER 14

A THEORETICAL FRAME FOR POST-CRISIS COMMUNICATION

Situational Crisis Communication Theory

W. Timothy Coombs
Eastern Illinois University

ABSTRACT

Post-crisis communication, what crisis managers say and do after a crisis, is a critical part of crisis management because it has significant ramifications for the organization's reputation. Unfortunately, this aspect of crisis management is underdeveloped in terms of theory. This chapter describes how attribution theory was used to develop situational crisis communication theory (SCCT), a theoretically based set of prescriptions to guide post-crisis communication. The application of attribution theory to post-crisis communication is examined through the exploration of SCCT.

Every organization faces the possibility of a crisis so they must be prepared for post-crisis communication—what an organization says and does after a crisis hits. One important goal of post-crisis communication is to protect

Attribution Theory in the Organizational Sciences, pages 275–296
Copyright © 2004 by Information Age Publishing

the organization's reputation (Dilenschneider, 2000). The time-pressured nature of post-crisis communication increases the chance for error. As is the case for the larger crisis management framework, preparation is the key to success (Barton, 2001; Mitroff & Anagnos, 2001). Crisis managers would benefit from detailed guidelines for post-crisis communication. While much has been written about post-crisis communication, there is a dearth of theory-driven recommendations for how post-crisis communication can serve to protect the organization's reputation (Seeger, Sellnow, & Ulmer, 1998, 2001).

Situational crisis communication theory (SCCT) has been developed as one option for filling this theoretical void (Coombs & Holladay, 2002). SCCT uses attribution theory as a framework for modeling the crisis situation and prescribing the selection of crisis response strategies. The purpose of this chapter is to detail how attribution theory was central to developing a theoretically based approach to post-crisis communication and to review the successes and failures associated with adapting attribution theory to post-crisis communication.

CHALLENGES FACING POST-CRISIS COMMUNICATION

An organization faces a variety of challenges during a crisis (Barton, 2001). One challenge is to protect/rebuild the organization's reputation. This section reviews the need to protect the organizational reputation and challenges to creating a systematic approach that utilizes post-crisis communication to protect reputational assets.

Reputation: Importance to Organizations

A reputation is how well or poorly stakeholders perceive an organization to be meeting stakeholder expectations (Bromley, 2000; Wartick, 2003). A reputation is evaluative; stakeholders make judgments about an organization being good or bad (Denbow & Culbertson, 1985; Wartick, 2003; Zyglidopoulos, 2003). Reputations are constructed from information stakeholders receive about the organization (Bradford & Garrett, 1995). That information can be derived directly from experience with the organization or indirectly from the news media or other secondhand sources of information (Mahon & Wartick, 2003). A reputation is widely accepted as a valued organizational resource because it can affect stock prices, recruitment, and even sales. Organizations exert considerable effort and resources to build favorable reputations (Fombrun, 1996; Wartick, 2003).

Hence, it is important to protect reputations from the threats posed by crises (Barton, 2001).

Post-Crisis Communication's Potential Contribution

Post-crisis communication, what an organization says and does after a crisis occurs, is perfectly suited to protecting an organization's reputational assets. Communication research has shown that words and actions can be used to address reputation-related concerns (Benoit, 1995; Benson, 1988; Coombs, 1995). To maximize its application, post-crisis communication needs to (1) identify a list of potential crisis situations, (2) articulate a list of possible crisis response strategies, and (3) develop a system for selecting the crisis response strategy that will maximize reputational protection in a given crisis—match the crisis response to the crisis situation (Benson, 1988; Coombs, 1995).

Researchers have touched on all three points over the past decade but have failed to provide an integration of the three. A variety of lists of crisis response strategies (e.g., Benoit, 1995; Marcus & Goodman, 1991; Siomkos & Shrivastava, 1993) and different types of crisis situations (e.g., Lerbinger, 1997; Mitroff, Harrington, & Gai, 1996; Pearson & Mitroff, 1993) have existed for years. However, the two lists generally were developed independently and were difficult to integrate due to their differing underlying conceptualizations (Coombs, 1995). One exception was Bradford and Garrett's (1995) Corporate Communicative Response Model. However, it was limited to crises involving charges of unethical behavior. Overall, the lists/taxonomies of crisis response strategies and crisis situations could not be used to match responses to the situation because the lists had different conceptual foundations and/or lack congruence. The potential of post-crisis communication's contributions were underdeveloped.

ATTRIBUTION THEORY: THE NECESSARY LINK

A crisis is an unexpected event that threatens to disrupt the operations of an organization and has negative effects on the organization (Coombs, 1999). Examples of crises include workplace violence, industrial accidents, and product recall/harm. A crisis is exactly the type of event that should trigger causal attributions because it is both unexpected (cannot be predicted) and unpleasant (has negative consequences for an organization and/or its stakeholders). Hence there is a natural connection between attribution theory and crisis management. Numerous management studies used attribution theory in their discussions of phenomena related to crisis communication, including charges of unethical behavior (Bradford & Garrett, 1995) and product recalls (Folkes, 1984; Griffin, Babin, & Attaway, 1991).

Attributions about causes do matter because they have affective and behavioral consequences (Weiner, 1985). Attributions can lead to anger or sympathy for a person. Strong attributions of personal responsibility for an event lead to greater feelings of anger and a more negative view of the target person (Wiener, Amirkan, Folkes, & Verette, 1987). Conversely, strong attributions of external control and weak attributions of personal control for an event lead to feelings of sympathy for the target person (Weiner, 1996). In turn, the negative emotions can affect how people interact with the target person (Weiner, 1985). It follows that attributions about crisis causality will affect perceptions of the organization (reputation), emotions toward the organization, and future interactions with the organization (behavioral intentions) (Coombs, 1995). Attribution theory provided a logical place to find a bridge for the gap between crisis response strategies and crisis situations.

Responsibility: The Lynch Pin

An unexpected, negative event leads people to search for attributions—to understand the cause of the event. People want to determine "why" something happened and will make attributions of the causality of an event based upon limited information (Bradford & Garrett, 1995; Kelley & Michela, 1980; Weiner, 1985; Weiner, Perry, & Magnusson, 1988). The basic choice in causality is between the person (internal) and the situation (external). Four causal dimensions have been identified as guiding attributions: stability, external control, personal control, and locus of causality. Stability refers to whether the cause of the event is permanent or changes over time. External control assesses whether or not some external agent could control the event. Personal control indicates whether or not the event was controllable by the person involved in the event (attributor). Locus of causality reflects if the cause of the event is something about the attributor or about the situation surrounding the event (McAuley, Duncan, & Russell, 1992; Russell, 1982; Wilson, Cruz, Marshall, & Rao, 1993).

Research has shown a significant overlap between the personal control and locus of causality dimensions (McAuley et al., 1992; Wilson et al., 1993). Both dimensions indicate intentionality of an event. Strong personal control and a high locus of causality both suggest that the attributor acted intentionally. McAuley and colleagues (1992) found evidence of the overlap but argued their analysis also showed the two dimensions could stand independently. Wilson and colleagues (1993) reported the overlap and suggested the use of just three dimensions: stability, external control, and locus/personal control (a combination of locus of causality and personal control).

In addition to causal dimensions, Kelley's (1972) idea of covariation posits three additional factors to consider when making attributions: consensus, consistency, and distinctiveness. Consensus involves comparison to relevant others. The target person is compared to see if they behaved similar to other people in the same situation. Would comparable people perform similar behaviors in the same situation? High consensus suggests the target person acted as others would act and attributions of causality to the person should be low. Consistency examines how a person performs a given task in the same situation over time. Does the target person engage in the behavior repeatedly? High consistency would increase attributions of causality to the target person. Distinctiveness considers how people behave across a variety of situations. Does the person act the same way in a variety of contexts? There is low distinctiveness when a person acts similarly across situations. Low distinctiveness should lead to stronger attributions of causality to the target person (Martinko, Douglas, Ford, & Gundlach, 2004).

Causality can lead to the assignment of responsibility, the judgment that someone should or ought to have done something else (Weiner, 1995). We are more likely to hold a person responsible when the locus/personal control is high, external control is low, and an event is stable. External control can be a source of mitigating circumstances, reasons the person could not control the event. In 2003, for instance, a man in Erie, Pennsylvania, was threatened with death if he did not help in a robbery. While robbery is typically a matter of internal locus/personal control, in this case the threat was an external control that would lessen perceptions of responsibility.

The same dynamic for assessing causality and responsibility should hold true for organizations. Stakeholders should assign greater crisis responsibility (consider the organization as responsible) for a crisis when the crisis situation is controllable by the organization, there are no external factors (mitigating circumstances), and the crisis is one in a series of problems (Coombs, 1995). A negative relationship is expected between crisis responsibility and the organizational reputation. The stronger the perceptions of crisis responsibility, the lower will be stakeholders' perceptions of the reputation—the more a crisis threatens the reputation. Crisis responsibility offers a potential framework for organizing the crisis situations, crisis response strategies, and the system for matching the two.

BASICS OF SCCT

A theory is a systematic view of a phenomenon that specifies relationships between variables for the purpose of explanation, prediction, and control (Neuliep, 1996). SCCT is a systematic view of post-crisis communication that identifies and specifies the relationship between key variables. SCCT explains how stakeholders view the crisis situation and the crisis response strategies. Crisis managers can use SCCT to predict how stakeholders will perceive a crisis situation and how they might perceive specific crisis response strategies. In turn, SCCT allows an organization some control over the reputational damage from a crisis. Crisis managers can select crisis response strategies that will maximize their potential to protect the reputation. SCCT demonstrates the basic characteristics of a theory.

SCCT uses crisis responsibility to articulate its list/taxonomy of crisis response strategies, organize its list of crisis situations, and develop a process of selecting the crisis response strategies that should maximize reputational protection. This section outlines the basic elements of SCCT and then presents the model that underlies the theory.

Crisis Response Strategies

Crisis response strategies are what crisis managers say and do after a crisis occurs. SCCT uses crisis responsibility as a central feature in developing its crisis response strategy list. Crisis managers can choose between three general response options: deny, diminish, or repair. The crisis response options indicate the primary communicative goal, the amount of responsibility an organization seems to accept, and the amount of aid it seems to provide for the crisis victims. In the deny response, the communicative goal is to prove no crisis occurred or that the organization has no responsibility for the crisis. No crisis or crisis responsibility means there is no reputational threat. There will be no attributions of crisis responsibility if the organization either has no crisis or is unconnected to the crisis. With the deny posture, no responsibility is accepted and no victims are acknowledged. In the diminish response, the communicative goal is to establish limited responsibility for the crisis and/or have stakeholders perceive the crisis damage as minor. The crisis manager is trying to affect how stakeholders view and make attributions about the crisis. The smaller the level of crisis responsibility or damage from the crisis, the less the reputational threat posed by the crisis. Limited acknowledgment of victims is provided. In the repair response, the communicative goal is to provide positive information about the organization to offset the negative information generated by the crisis. Organizations are perceived as taking responsibility for the crisis and concentrating their responses on victim concerns. One way to combat the reputational threat is by offering stakeholders reasons to think well of the organization. Crisis managers are using impression management in attempts to construct a more favorable reputation (Bradford & Garrett, 1995; Coombs, 1995).

For SCCT, a list of crisis response strategies was compiled from the existing literature. Potential crisis response strategies were gleaned from a wide range of works related to crisis communication and impression management (e.g., Bolino & Turnley, 2003; Bradford & Garrett, 1995; Hearit, 1996; Marcus & Goodman, 1991; Mohamed, Gardner, & Paolillo, 1999). Crisis researchers have drawn upon corporate apologia and account-giving to develop lists of communicative responses. Corporate apologia is the public discourse used to protect an organization's public persona/reputation when it has been attacked in some manner (Dionisopolous & Vibbert, 1988). A crisis is a form of attack on the organizational reputation and corporate apologia can be offered as a defense (Hearit, 1994, 1995a, 1995b, 2001). An account is a message designed to explain why a person committed an unseemly act. The account is also an attempt to influence how others perceive you/to shape your reputation (Benoit, 1995). Corporate apologia and accounts share the use of communication in attempts to affect perceptions of a reputation.

The final list of crisis response strategies is presented in Table 14.1 and organized according to the deny, diminish, or repair response options. The strategies reflect varying degrees of accepting responsibility and addressing victim concerns.

Table 14.1.   Crisis Response Strategies to Response Options

Deny Strategies: Attempts to completely disconnect the organization from the crisis.

Attack the accuser: crisis manager confronts the person or group claiming something is wrong with the organization.
Denial: crisis manager claims that there is no crisis.
Scapegoat: crisis manager blames some person or group outside of the organization for the crisis.

Diminish Strategies: Attempts to minimize responsibility and/or damage from the crisis.

Excuse: crisis manager minimizes organizational responsibility by denying intent to do harm and/or claiming inability to control the events that triggered the crisis.
Justification: crisis manager minimizes the perceived damage caused by the crisis.

Repair Posture: Attempts to provide positive information about the organization.

Ingratiation: crisis manager praises stakeholders.
Bolstering: reminds stakeholders of past good works by the organization.
Compensation: crisis manager offers money or other gifts to victims.
Apology: crisis manager indicates the organization takes full responsibility for the crisis and asks stakeholders for forgiveness.

Crisis Situations

Crisis situations have been conceptualized in a variety of ways. SCCT uses reputational threat as the focal point for organizing and defining the crisis situation. While a crisis situation can vary by other factors, reputational threat provides a connecting point between the crisis response strategies and crisis situations through responsibility. Crisis responsibility is used as the key determinant of the reputational threat posed by a crisis situation.

Evaluation of the reputational threat created by a crisis situation is a two-step process. In the first step, the crisis manager determines the crisis type. A crisis type is a category for a crisis that frames how people interpret the crisis event. For instance, was the crisis event management misconduct or an accident? The crisis management literature was reviewed and a list of 10 crisis types was identified for SCCT. A crisis type was included on the list only if it had been discussed by two or more crisis experts (Coombs, 1999b). Table 14.2 lists and defines the 10 crisis types.

Table 14.2. Crisis Types by Reputational Threat Level

Crisis Types That Generate a Extreme Reputation Threat

Human-error accidents: human error causes an industrial accident.
Human-error recalls: human error causes a product to be recalled.
Organizational misdeed: management knowingly places stakeholders at risk or violates the law. Includes three variations: (1) with no injuries: stakeholders are deceived without injury; (2) management misconduct: laws or regulations are violated by management; and (3) with injuries: stakeholders are placed at risk by management and injuries occur.

Crisis Types That Generate a Moderate Reputation Threat

Challenges: stakeholders claim an organization is operating in an inappropriate manner.
Technical-error accidents: a technology or equipment failure causes an industrial accident.
Technical-error recalls: a technology or equipment failure causes a product to be recalled.

Crisis Types That Generate a Mild Reputational Threat

Natural disaster: acts of nature that damage an organization such as an earthquake.
Rumors: false and damaging information about an organization is being circulated.
Workplace violence: current or former employee attacks current employees onsite.
Product tampering/malevolence: external agent causes damage to an organization.

Different crisis types will generate different attributions of crisis responsibility. Research has grouped the 10 crisis types into three threat levels: mild, moderate, and extreme. The mild level produces very weak perceptions of crisis responsibility and includes the natural disaster, product tampering, workplace violence, and rumor crisis types. The moderate level produces low-level perceptions of crisis responsibility and includes the technical-error accident, technical-error product recall, and challenge crisis types. The extreme level produces very strong perceptions of crisis responsibility and includes the human-error accident, human-error product recall, and organizational misdeeds (Coombs & Holladay, 2002).

Perceptions of crisis responsibility are largely a function of the attributions of personal control/locus of control. As Weiner (1986) has noted, the relevance of the dimensions of attribution are dependent on the domain of interest. External control was found to explain little variance in crisis responsibility and dropped from SCCT (Coombs, 1998). In the Coombs and Holladay (2002) study that produced groupings of crisis types, personal control/locus was measured using modified items from the personal control and locus of causality dimensions of McAuley and colleagues' (1992) Revised Causal Dimension Scale (CDSII; Coombs & Holladay, 2002). Crisis responsibility was measured using revised items from Griffin and colleagues' (1992) three-item scale for blame. A cluster analysis based on crisis responsibility was used to determine the levels of reputational threat (Coombs & Holladay, 2002).

The second step in assessing the reputational threat of the crisis situation involves crisis managers adjusting the reputational threat if any of three intensifiers are present: (1) crisis history, (2) relationship history, and (3) severity of the crisis. The intensifiers were drawn from attribution theory.

Crisis history is whether or not an organization has had a history of similar crises. Relationship history is how well or how poorly an organization has treated key stakeholders in the past. Crisis and relationship history represent the stability dimension of causality. A history of crises indicates the current crisis is part of a stable pattern of behavior rather than an aberration, thus increasing perceptions of crisis responsibility. A crisis becomes just one more in a pattern of negative behaviors found in the unfavorable relationships history, thus intensifying perceptions of crisis responsibility. The news media frequently report if an organization has had a similar crisis. People often have a general idea of how well or poorly an organization has treated its stakeholders. Again, media coverage also provides a picture of how an organization interacts with stakeholders (Carrol & McComb, 2003). An unfavorable relationship history will intensify the reputational threat. Crisis history and relationship history have an indirect effect on organizational reputation through crisis responsibility.

Crisis and relationship history are also related to the causal antecedents, information that can shape attributions, found in Kelley's (1972) idea of covariation. Crisis history is a form of consistency, the organization has had similar crises—has acted the same way at other times. Relationship history is a form of distinctiveness, the organization has problems in a variety of contexts (Weiner, 1996). Distinctiveness and consistency are part of covariation. Covariation argues that people look for an association between a cause and an effect across a variety of conditions (Martinko et al., 2004).

Consensus, the third element of covariation, has yet to be integrated into SCCT. An organization with a history of crises or an unfavorable relationship history will create stronger attribution of personal control. In turn, the organization will experience stronger perceptions of crisis responsibility and a greater reputational threat exists. As in other attribution situations, stakeholders often lack information about distinctiveness and consistency in crisis situations.

Severity is related to the attribution theory concept of extremeness. The more intense the effect of a crisis, the more likely people are to attribute personal control to an organization. In turn, the increased personal control will intensify perceptions of crisis responsibility and the reputation threat presented by the crisis situation (Kelley & Michela, 1980).

The three intensifiers have a direct effect on crisis responsibility as well. In impression management, people try to control the information people receive about them in order to control the impressions others form about them (Bradford & Garrett, 1994; Caillouet & Allen, 1996; Giacalone & Pollard, 1987). Reputation management is impression management for an organization; managers try to control the information stakeholders receive about an organization to shape the stakeholders' view of the organization's reputation (Allen & Caillouet, 1994; Bradford & Garrett, 1995). Crisis history, unfavorable relationship history, and severity are all bits of information about an organization that become salient to reputation formation following a crisis. Severity is the amount of harm done by or damage inflicted by a crisis. Damage can be human, property, or financial. Severe damage, a history of crises, and an unfavorable relationship history are all very negative information that should lead to negative reputations/impressions. Figure 14.1 illustrates the relationship between personal control, crisis responsibility, organizational reputation, and the intensifiers.

The presence of any one or combination of the intensifiers should alter the reputational threat posed by the crisis situation. Mild crises become perceived as moderate crises and moderate crises become perceived as extreme crises when one or more of the intensifiers is present. Intensifiers increase the reputational threat posed by a crisis situation. In summary, the assessment of the crisis situation involves identifying the crisis type to

Figure 14.1.   Situational crisis communication theory model.

determine the initial reputational threat from the crisis. The initial threat is adjusted upward if the crisis damage is severe, if there were past crises, and/or if past relationships with stakeholders were unfavorable. The crisis situation is a constellation of crisis type, crisis history, relationship history, and severity. When taken together, these factors provide a picture of how stakeholders are likely to perceive crisis responsibility for the crisis event and the reputational threat posed by the crisis situation.

Perceptions of crisis responsibility should shape the affective reaction to the crisis. High levels of crisis responsibility should induce anger. "The organization should have done something to prevent the crisis." Very low levels of crisis responsibility could create sympathy as the organization is seen as a victim of the crisis. "It is too bad this crisis happened." The emotions triggered by the crisis can affect both the organization's reputation and behavioral intentions. Stakeholders are more likely to view the organization negatively if they are angry and less likely to interact with the organization in the future.

Prescriptions from SCCT: Matching the Crisis Situation to Crisis Response Strategies

Every crisis response needs to begin with instructing and adjusting information before utilizing reputation-building strategies. Instructing information tells stakeholders what happened, how the crisis might affect them, and what they can do to protect themselves physically from the crisis such as shelter-in-place or evacuate after a chemical release. Adjusting information helps stakeholders to cope psychologically with the crisis (Sturges, 1994). Compassion, offering words of concern, and corrective action, telling stakeholders what is being done to prevent a repeat of the crisis, are the core of adjustment. The ambiguity created by a crisis creates a need among stakeholders for information. A critical piece of information is what the organization is doing to protect stakeholders from similar crises in the future—what corrective actions are being taken. Corrective actions reassure stakeholders that they are safe, which should serve to reduce their psychological stress (Sellnow, Ulmer, & Snider, 1998). Stakeholders expect that the organization will acknowledge the victims in some fashion, typically with an expression of concern (Patel & Reinsch, 2003). This recommendation to express sympathy or concern for victims is derived from the need for adjusting information. All post-crisis communication should begin with instructing and adjusting information (Sturges, 1994). Once instructing and adjusting information are provided, crisis managers can address reputational concerns.

Rumors and challenges present unique situations for crisis managers because each involve the possibility of there being no crisis. By definition, a rumor is untrue and crisis managers may seek to disprove it or simply not respond. The deny crisis response strategies are perfectly suited to a rumor. Challenges question an organization's behavior on moral or ethical grounds. One option is to deny there are any ethical or moral violations (Bradford & Garrett, 1995; Fearn-Banks, 1996).

For a crisis that generates a mild reputational threat, crisis managers need only offer instructing and adjusting information. Stakeholders already perceive very low crisis responsibility, so additional reminders are unnecessary. A crisis that generates a moderate reputational threat requires the diminish crisis response strategies. The crisis manager tries to reinforce the low-level crisis attributions with excuses and/or justifications. A crisis that creates an extreme reputational threat demands the repair crisis response strategies. Image restoration theory places a heavy emphasis on apology, publicly accepting responsibility for the crisis (Benoit, 1995; Tyler, 1997). However, the legal and financial burdens created by an apology may be too much for an organization to bear. As a result, crisis managers may rely on compensation, bolstering, ingratiation, or some combination of the three. Bolstering and ingratiation are weak if no compassion is offered in the adjusting information because neither bolstering nor ingratiation offer much concern for victims or address issues of responsibility. Hence, it is best to use compassion, apology, and/or compensation with bolstering and/or ingratiation when the crisis represents an extreme reputational threat. Table 14.3 provides a summary of the crisis response recommendations provided by SCCT.

Table 14.3.   Crisis Response Strategy Recommendations

A. Rumor: use any of the denial strategies.

B. Challenge: use denial strategies when attack is unfair.

C. Mild Reputational Threat Crisis Situations (Natural Disaster, Workplace Violence, and Product Tampering with no Intensifiers): provide instructing and adjusting information.

D. Moderate Reputational Threat Crisis Situations (Natural Disaster, Workplace Violence, and Product Tampering with Intensifiers and Technical-Error Product Recall, Technical-Error Accident, and Challenge with no Intensifiers).

 1. Use excuse crisis response strategy to reinforce the involuntary nature of the crisis.

 2. Use justification strategy when stakeholders might misperceive the damage created by the crisis.

E. Extreme Reputational Threat Crisis Situation (Technical-Error Product Recall, Technical-Error Accident, and Challenge with Intensifiers and Human-Error Product Recall, Human-Error Accident, or Organizational Misdeed).

 1. Use compensation to help rebuild the reputation and limit liability.

 2. Use compensation and apology when organization is willing and able to accept the legal and financial liabilities.

Table 14.3.   Crisis Response Strategy Recommendations (Cont.)

3. Use bolstering and/or ingratiation if liability risks are great and compassion has been expressed.

F. Any Crisis Situation

  1. Use bolstering as a supplemental strategy when the organization has a history of good work.

  2. Use ingratiation as a supplemental strategy.

Scope of SCCT

SCCT does have a limited scope for application. SCCT can be used by organizations when they face any of its 10 crisis types. SCCT is not meant to apply to individuals such as politicians or celebrities. The narrow focus is warranted to accommodate the unique elements of corporate crises. For instance, organizational responses are often constrained by financial and legal concerns (Coombs, 2002). Attempts to create post-crisis communication principles that are too generic have been criticized for not recognizing the realities of corporate life (Tyler, 1997).

Summary of SCCT Relationships

A review of the relations presented in Figure 14.1 will help to clarify SCCT. The trigger is a crisis event. Stakeholders will make attributions of personal control—the degree to which they believe the organization could have controlled crisis events. The attributions of personal control are positively related to perceptions of crisis responsibility. Severity, a history of crises, and poor relationships with stakeholders (relationship history) will intensify perceptions of crisis responsibility. Crisis responsibility is negatively related to organizational reputation. Moreover, a history of crises and poor relationships with stakeholders has a direct effect on organizational reputation. The organizational reputation will affect the behavioral intentions of the stakeholders. Crisis response strategies can be used to alter perceptions of crisis responsibility and/or reputational damage.

LESSONS FROM ADAPTING ATTRIBUTION THEORY TO SCCT

A number of principles designed for individuals, such as apologia and image restoration, have been adapted for post-crisis communication (e.g., Benoit, 1995; Hearit, 2001). The fit for these concepts is rarely exact as principles designed for individuals may not have equivalent effects when applied to organizations. This section reviews the successes and failures of translating attribution theory into post-crisis communication through SCCT.

Successful Applications

Attribution theory provided a framework for developing the variables used to assess the reputational threat of a crisis. Personal control, an attribution theory concept, is a foundational element of SCCT. Research supports the positive relationship between personal control and crisis responsibility (Coombs, 1998, 1999a; Coombs & Holladay, 2001, 2002; Coombs & Schmidt, 2000; Nerb & Spada, 1997). Crisis responsibility has proven to be negatively related to organizational reputation (Coombs, 1998, 1999a; Coombs & Holladay, 2001, 2002; Coombs & Schmidt, 2000). Personal control does help to predict crisis responsibility and establish the initial crisis threat posed by the crisis situation. Each crisis type generates predictable levels of personal control and crisis responsibility, thus providing the initial assessment of the reputational threat.

The second step in assessing the reputational threat of a crisis involves crisis history and relationship history (Coombs & Holladay, 2001). Crisis and relationship history reflect the attribution theory concepts of stability and causal antecedents. Crisis and relationship history have proven to intensify the reputational threat of a crisis situation. They are useful in assessing the crisis situation. As with more general attributions, distinctiveness/relationship history was the more important of the two variables as it explained more variance (Coombs, 1998; Coombs & Holladay, 1996, 2001). It should be noted that the exact threshold level for crisis history and performance history have yet to be established. Current research suggests one past crisis might be enough to trigger consistency and information about one additional domain of negative behavior might be enough to engage distinctiveness. More research is necessary to more fully establish the thresholds. Without a means of predicting the reputational threat of a crisis, SCCT would lack prescriptive value. SCCT drew upon previous research to locate key features of the crisis situation that a crisis manager can use to anticipate the reputational threat of the current crisis. During a crisis there is not time to measure how stakeholders perceive the crisis. By identifying critical aspects of the crisis situation (crisis type and intensifiers), crisis managers can anticipate the reputational threat of the crisis and, in turn, select the appropriate crisis response strategy(ies). Attribution theory provided the framework for identifying the variables used to assess the reputational threat inherent in a crisis situation.

The predicted relationships between crisis responsibility–behavioral intention and organizational reputation–behavioral intention were found as well (Coombs & Holladay, 2001; Coombs & Schmidt, 2000). Perceptions of the crisis and the organization in crisis do seem to impact future interactions. The successful applications of attribution theory have been tested across a wide variety of crisis situations (Coombs & Holladay, 2001, 2002).

Failed Applications

The attribution concepts of external control and severity have yet to be proven useful in post-crisis communication. However, external control did provide a foundation for categorizing crisis types. External control and personal control were used to create a 2 x 2 matrix for categorizing crisis types (Coombs et al., 1997). However, external control failed to correlate with other important variables in SCCT and explained little variance for crisis responsibility, organizational reputation, or behavioral intentions (Coombs, 1998; Coombs & Holladay, 1996). The crises in the mild reputational threat condition also have high attributions of external control. However, external control does not contribute significantly to explaining the variance related to crisis responsibility or organizational reputation.

It could be that stakeholder attributions about organizational actions focus on personal control with little effect from external control. External control is not a relevant mitigating factor for some crisis types. This could be related to the general notion that stakeholders dislike organizational attempts to shift blame. External control may be useful once emotion is introduced into SCCT but has yet to provide much utility to post-crisis communication. Another possibility is that external control was abandoned too soon. The role of external control as a mitigating factor may be relevant to the effects of various crisis response strategies. Do the crisis response strategies help to protect the organization's reputation by affecting attributions of external control? This remains a possibility but stakeholders' general negative reactions to shifting blame may be the more important factor.

Severity, the amount of damage inflicted by a crisis, has not produced consistent findings. As a form of extremity, attributions of personal control should increase as severity intensifies. Empirical tests have not found a consistent correlation between severity and personal control or crisis responsibility (Coombs & Holladay, 2004). The problem could be an issue of operationalizing severity or the effect of severity could vary according to the crisis type. Further testing is required to determine if severity can demonstrate a predictable relationship to personal control or crisis responsibility. One possibility is to follow Laufer's lead (Laufer & Gillespie, 2004) and conceptualize severity as personal vulnerability to the crisis. Severity could be the fear that a similar crisis could affect the stakeholder rather than some level of damage.

The discounting principle argues that a given cause for a crisis could be discounted if other plausible causes are offered. On the surface, the discounting principle holds promise for post-crisis communication. An organization could offer a variety of possible causes suggesting limited organizational control in order to blunt attributions of personal control and crisis responsibility. For instance, an organization could list a variety of technical reasons for an industrial accident to prevent stakeholders from assuming the accident was human error. Thinking about technical errors rather than human error would lessen the reputational threat of the crisis (Coombs & Holladay, 2002). However, a basic rule in crisis management, "never speculate during a crisis," works against employing the discounting principle. This would include offering possible causes for a crisis. Avoiding speculation precludes offering possible causes for a crisis. Conjecture is dangerous because if the organization's speculation is proven wrong, it would appear that they were either being deceptive or lacked knowledge (Coombs, 1999). Neither is the type of perception an organization would like to cultivate among stakeholders. How the discounting principle can be used without running the risks of speculation is worthy of further consideration.

Tests of Prescriptions

Crisis response strategy selection prescriptions have received minimal testing. The first test of the prescriptions found support for a diminish strategy in a moderate crisis threat crisis situation (technical-error accident with no intensifiers) and for a repair strategy in an extreme crisis threat crisis (organizational misdeed). For each crisis type, the matched crisis response condition produced less organizational reputation damage than either the no response or mismatched conditions (Coombs & Holladay, 1996). The value of the repair strategy in an extreme crisis threat crisis (organizational misdeed) was replicated in a later study (Coombs & Schmidt, 2000).

A third study demonstrated that instructing information provided sufficient reputational protection for mild crisis threats (natural disaster and product tampering crisis types with no intensifiers). The tests indicated the diminish and repair strategies provided no additional reputational benefits beyond those already created by the instructing information (Coombs & Holladay, 2004). Further testing is needed to determine the precise nature of adjusting information in the crisis response. Additional testing also is necessary to examine a wider array of the crisis response strategy prescriptions. Moreover, the research should determine if the diminish strategies have the desired effect of reducing perceptions of crisis responsibility. Previous research failed to check for this effect and only examined the impact of the diminish crisis response strategy on organizational reputation (Coombs & Holladay, 1996). Thus far, testing the prescriptive recommendations of SCCT has been limited and thin.

FUTURE DEVELOPMENT OF SCCT
USING ATTRIBUTION THEORY

SCCT is in its early stages of testing and development. Additional ideas derived from attribution theory could and should be integrated into SCCT and tested. Three promising ideas are emotion, the fundamental attribution error, and consensus.

Emotion and SCCT

Thus far, SCCT has yet to examine the role of emotions in the crisis management process. Attribution theory holds that emotions can mediate between perceptions of responsibility and behavioral intentions (Weiner, 1995). A crisis can create either anger or sympathy. Stakeholders may become angry with an organization when they believe the organization could and should have prevented the crisis. Crises that generate strong perceptions of crisis responsibility should also create anger. External control might be relevant here as it could be a mitigating factor for anger. More research is needed to identify how anger fits into post-crisis communication, its effects on behavioral intentions, and the role of mitigating factors. Included in mitigating factors would be how crisis response strategies could be used to highlight or to introduce mitigating factors for anger.

Sympathy for the organization is another possible emotion created by a crisis. Crises that pose mild reputation threats should evoke sympathy for the organization. The organization is a victim of external forces rather than its own actions. How does sympathy fit into SCCT and post-crisis communication? Additional research is needed to determine how crisis response strategies can be used to intensify feelings of sympathy and the effect of sympathy on behavioral intentions. The greatest weakness in SCCT has been the failure to consider and to integrate emotion into its conceptual framework. The model of SCCT presented in this chapter represents the first attempt to integrate affect into the theory. Refer to Figure 14.1 for the model.

Fundamental Attribution Error

There has been no systematic effort to determine if stakeholders fall victim to the fundamental attribution error. Accident crisis type permits an exploration of fundamental attribution error. An accident crisis type can have either a human error (high personal control) or technical error (low personal control) cause. Managers tend to attribute accidents to human error (Perrow, 1999; Reason, 1999). Some tentative results suggest that stakeholders do not follow the same pattern. They do not assume the crisis is human error when no cause is given. Fundamental attribution error may be a function of other organizational perceptions such as organizational cynicism and performance history. The role of fundamental attribution error in SCCT should be explored more fully.

Consensus

SCCT has yet to adapt and evaluate consensus as a factor, one of the three variables in covariation. Consensus is a social comparison. If stakeholders could compare how the current crises match to how other organizations perform in the same area, such as chemical accidents or recalls, this could alter the perceptions of crisis responsibility. If many organizations have been hit by similar crises, the attributions of crisis responsibility should be lessened. The crisis would be seen as an industry rather than an organization-specific problem. If the crisis is rare in the industry, the perceptions of crisis responsibility should be stronger. The crisis would seem organization-specific. An organization can supply contextual information that would help people assess consensus such as other, similar crises in the industry. However, there is a danger that the organization might be trying to shift blame to the industry and this could be viewed negatively. The parameters of supplying consensus-related information in the news media and its effects on crisis responsibility and reputation have yet to be developed.

DISCUSSION

SCCT was designed to be applied to a variety of organizational crises rather than be a broad theory applied in any crisis setting. Attribution theory provided the foundation for integrating crisis response strategies and crisis situations into a coherent theory of post-crisis communication. Variables and explanatory mechanisms from attribution theory were used to construct the basic SCCT model by suggesting how the variables could be related. The direct translation of attribution theory into post-crisis communication has proven somewhat problematic. External control seems to play a less important role in SCCT than in attributions about individual actions. Organizations have to be careful not to be perceived as shifting the blame to external factors because stakeholders may react negatively to scapegoating. Stakeholders seem to expect organizations to face up to crises and not blame others for the situation. Crisis managers must exercise caution in how they present mitigating factors. This has implications for utilization of the discounting principle as well. More attention needs to be directed to if and how external control and the discounting principles can be integrated into SCCT.

Personal control, distinctiveness (relationship history), and consistency (crisis history) have proved to be valuable in predicting stakeholder perceptions of crisis responsibility and the reputation threat posed by a crisis. Predicting the reputational threat of a crisis is the central concern of SCCT. SCCT bases its recommendations for the selection of crisis response strategies on the reputation threat associated with a crisis. By understanding how elements of the crisis situation (personal control, distinctiveness, and consistency) might affect the reputational threat, a crisis manager has theoretically based rationales for selecting crisis response strategies designed to protect the organization's reputational assets. Attribution theory has been conceptually important to the development of a theory-based approach for post-crisis communication.

REFERENCES

Allen, M. W., & Caillouet, R. H. (1994). Legitimate endeavors: Impression management strategies used by an organization in crisis. *Communication Monographs, 61*, 44–62.

Barton, L. (2001). *Crisis in organizations II* (2nd ed.). Cincinnati, OH: College Divisions SouthWestern.

Benoit, W. L. (1995). *Accounts, excuses, and apologies: A theory of image restoration strategies.* Albany: State University of New York Press.

Benson, J. A. (1988). Crisis revisited: An analysis of the strategies used by Tylenol in the secondtampering episode. *Central States Speech Journal, 38*, 49–66.

Bitzer, L. F. (1968). The rhetorical situation. *Philosophy and Rhetoric, 1*, 165–168.

Bolino, M. C., & Turnley, W. H. (2003). More than one way to make an impression: Exploringprofiles of impression management. *Journal of Management, 29*, 141–160.

Bradford, J. L., & Garrett, D. E. (1995). The effectiveness of corporate communicative responsesto accusations of unethical behavior. *Journal of Business Ethics, 14*, 875–892.

Bromley, D. B. (2000). Psychological aspects of corporate identity, image and reputation. *Corporate Reputation Review, 3*, 240–252.

Coombs, W. T. (1995). Choosing the right words: The development of guidelines for the selection of the "appropriate" crisis response strategies. *Management Communication Quarterly, 8,* 447–476.

Coombs, W. T. (1998). An analytic framework for crisis situations: Better responses from a better understanding of the situation. *Journal of Public Relations Research, 10,* 177–191.

Coombs, W. T. (1999a). Information and compassion in crisis responses: A test of their effects. *Journal of Public Relations Research, 11,* 125–142.

Coombs, W. T. (1999b). *Ongoing crisis communication: Planning, managing, and responding.* Thousand Oaks, CA: Sage.

Coombs, W. T. (2002). *Further testing of the situational crisis communication theory: Anextended examination of crisis history as a modifier.* Paper presented at the annual meeting of the National Communication Association, New Orleans, LA.

Coombs, W. T., & Holladay, S. J. (1996). Communication and attributions in a crisis: Anexperimental study of crisis communication. *Journal of Public Relations Research, 8,* 279–295.

Coombs, W. T., & Holladay, S. J. (2001). An extended examination of the crisis situation: Afusion of the relational management and symbolic approaches. *Journal of Public Relations Research, 13,* 321–340.

Coombs, W. T., & Holladay, S. J. (2002). Helping crisis managers protect reputational assets:Initial tests of the situational crisis communication theory. *Management Communication Quarterly, 16,* 165–186.

Coombs, W. T., & Holladay, S. J. (2004). Reasoned action in crisis communication: Anattribution theory-based approach to crisis management. In D. P. Millar & R. L. Heath (Eds.), *Responding to crisis: A rhetorical approach to crisis communication* (pp. 95–115). Ordina, CA: Erlbaum.

Coombs, W. T., & Schmidt, L. (2000). An empirical analysis of image restoration: Texaco'sracism crisis. *Journal of Public Relations Research, 12,* 163–178.

Denbow, C. J., & Culbertson, H. M. (1985). Linking beliefs and diagnosing image. *PublicRelations Review, 11,* 29–37.

Dilenschneider, R. L. (2000). *The corporate communications bible: Everything you need toknow to become a public relations expert.* Beverly Hills, CA: New Millennium Press.

Dionisopolous, G. N., & Vibbert, S. L. (1988). CBS vs Mobil Oil: Charges of creativebookkeeping. In H. R. Ryan (Ed.), *Oratorical encounters: Selected studies and sources of 20th century political accusation and apologies* (pp. 214–252). Westport, CT: Greenwood Press.

Fearn-Banks, K. (1996). *Crisis communications: A casebook approach.* Mahwah, NJ: Erlbaum.

Fombrun, C. J. (1996). *Reputation: Realizing value from the corporate image.* Boston: HarvardBusiness School Press.

Fombrun, C., & Shanely, M. (1990). What's in a name? Reputation building and corporatestrategy. *Academy and Management Journal, 33,* 233–258.

Giacalone, R. A., & Pollard, H. G. (1987). The efficacy of accounts for a breach of confidence by management. *Journal of Business Ethics, 6,* 393–397.

Griffin, M., Babin, B. J., & Darden, W. R. (1992). Consumer assessments of responsibility forproduct-related injuries: The impact of regulations, warnings, and promotional policies. *Advances in Consumer Research, 19,* 870–877.

Hearit, K. M. (1994). Apologies and public relations crises at Chrysler, Toshiba, and Volvo, *Public Relations Review, 20,* 113–125.

Hearit, K. M. (1995a). From "we didn't it" to "it's not our fault": The use of apologia in publicrelations crises. In W. N. Elwood (Ed.), *Public relations inquiry as rhetorical criticism: Case studies of corporate discourse and social influence* (pp. 117–134). Westport, CT: Praeger.

Hearit, K. M. (1995b). "Mistakes were made": Organizations, apologia, and crises of sociallegitimacy. *Communication Studies, 46,* 1–17.

Hearit, K. M. (1996). The use of counter-attack in apologetic public relations crises: The case ofGeneral Motors vs. Dateline NBC. *Public Relations Review, 22,* 233–248.

Hearit, K. M. (2001). Corporate apologia: When an organization speaks in defense of itself. In R.L. Heath (Ed.), *Handbook of public relations* (pp. 501–511). Thousand Oaks, CA: Sage.

Herbig, P., Milewicz, J., & Golden, J. (1994). A model of reputation building and destruction. *Journal of Business Research, 31,* 23–31.

Kelley, H. H. (1972). Causal schema and the attribution process. In E. E. Jones, D. E. Kanouse,H. H. Kelley, R. E. Nisbett, S. Valins, & B. Wiener (Eds.), *Attributions: Perceiving the causes of behavior* (pp. 151–174). Morristown, NJ: General Learning Press.

Kelley, H. H., & Michela, J. L. (1980). Attribution theory and research. *Annual Review of Psychology,* 31, 457–501.

Laufer, D., & Gillespie, K. (2004). Differences in consumer attributions of blame between menand women: The role of perceived vulnerability and empathic concern. *Psychology and Marketing, 21,* 141–157.

Lerbinger, O. (1997). *The crisis manager: Facing risk and responsibility.* Mahwah, NJ: Erlbaum.

Mahon, J. F., & Wartick, S. L. (2003). Dealing with stakeholders: How reputation, credibility and framing influence the game. *Corporate Reputation Review, 6,* 19–35.

Marcus, A. A., & Goodman, R. S. (1991). Victims and shareholders: The dilemmas of presentingcorporate policy during a crisis. *Academy of Management Journal,* 34, 281–305.

Martinko, M. J., Douglas, S. C., Ford, R., & Gundlach, M. J. (2004). Dues paying: A theoreticalexplication and conceptual model. *Journal of Management, 30,* 49–69.

McAuley, E., Duncan, T. E., & Russell, D. W. (1992). Measuring causal attributions: The revised causal dimension scale (CDII). *Personality and Social Psychology Bulletin, 18,* 566–573.

McCroskey, J. C. (1966). *An introduction to rhetorical communication.* Englewood Cliffs, NJ: Prentice-Hall.

Metts, S., & Cupach, W. R. (1989). Situational influence on the use of remedial strategies inembarrassing predicaments. *Communication Monographs, 56,* 151–162.

Mitroff, I. I., & Anagnos, G. (2001). *Managing crises before they happen: What every executiveand manager needs to know about crisis management.* Chicago: AMACOM.

Mitroff, I. I., Harrington, K., & Gai, E. (1996). Thinking about the unthinkable. *Across the Board, 33*(9), 44–48.

Mohamed, A. A., Gardner, W. L., & Paolillo, J. G. P. (1999). A taxonomy of organizationalimpression management tactics. *ACR, 7,* 108–130.

Patel, A., & Reinsch, L. (2003). Companies can apologize: Corporate apologies and legal liability. *Business Communication Quarterly, 66,* 17–26.

Pearson, C. M., & Mitroff, I. I. (1993). From crisis prone to crisis prepared: A framework forcrisis management. *The Executive, 7,* 48–59.

Russell, D. (1982). The causal dimension scale: A measure of how individuals perceive causes. *Journal of Personality and Social Psychology, 42,* 1137–1145.

Seeger, M. W., Sellnow, T. L., & Ulmer, R. R. (1998). Communication, organization, and crisis. In M. E. Roloff (Ed.), *Communication Yearbook 21* (pp. 231–276). Thousand Oaks, CA: Sage.

Seeger, M. W., Sellnow, T. L., & Ulmer, R. R. (2001). Public relations and crisis communication: Organizing and chaos. In R. L. Heath (Ed.), *Handbook of public relations* (pp. 155–166). Thousand Oaks, CA: Sage.

Sellnow, T. L., Ulmer, R. R., & Snider, M. (1998). The compatibility of corrective action inorganizational crisis communication. *Communication Quarterly, 46,* 60–74.

Sturges, D. L. (1994). Communicating through crisis: A strategy for organizational survival. *Management Communication Quarterly, 7,* 297–316.

Tyler, L. (1997). Liability means never being able to say you're sorry: Corporate guilt, legal constraints, and defensiveness in corporate communication. *Management Communication Quarterly, 11,* 51–73.

Wartick, S. L. (2003). Measuring corporate reputation: Design and data. *Business and Society,* 41, 371–392.

Weiner, B. (1985). An attributional theory of achievement motivation and emotion. *Psychology Review, 92,* 548–573.

Wiener, B. (1986). *An attributional theory of motivation and emotion.* New York: Springer-Verlag.

Weiner, B. (1996). *Judgments of responsibility: A foundation for a theory of social conduct.* New York: Guilford Press.

Weiner, B., Amirkan, J., Folkes, V.S., & Verette, J.A. (1987). An attribution analysis of excuse giving: Studies of a naive theory of emotion. *Journal of Personality and Social Psychology, 53,* 316–324.

Weiner, B., Perry, R.P., & Magnusson, J. (1988). An attribution analysis of reactions to stigmas. *Journal of Personality and Social Psychology, 55,* 738–748.

Wilson, S. R., Cruz, M.G., Marshall, L. J., & Rao, N. (1993). An attribution analysis of compliance-gaining interactions. *Communication Monographs, 60,* 352–372.

Winkleman, R. (1999). The right stuff. *Chief Executive, 143*(4), 80–81.

Zyglidopoulos, S. C. (2003). The issue life-cycle: Implications for reputation for social performance and organizational legitimacy. *Corporate Reputation Review, 6,* 70–81.

CHAPTER 15

PARTING THOUGHTS

Current Issues and Future Directions

Mark J. Martinko
Florida State University

Throughout the symposium and the preceding chapters, a number of issues emerged that represent both challenges and opportunities for the development and application of attribution theory. Some of these issues emerged in the course of the discussion of specific papers and others emerged when the participants were asked to identify the issues that were most salient to them. The purpose of this final chapter is to summarize the issues that were discussed the most and use these issues to help identify future directions and challenges for the evolution and application of attribution theory.

BASIC ASSUMPTIONS

Before discussing these issues and future directions, it is important to state some of my basic assumptions regarding the role and importance of attribution theory as an explanatory construct for organizational behavior. At the onset, it should be noted that many of the conference participants, but not all of them, share a similar set of assumptions.

Attribution Theory in the Organizational Sciences, pages 297–305

My most basic assumption is that attributional processes are an integral part of goal-oriented behaviors. Almost all behavioral and organizational theorists accept the basic Skinnerian premise that behavior is a function of its consequences; that people seek to maximize their rewards and minimize their pain. Thus, individuals gain both personal and organizational control to the extent that they are able to manage and control their reward systems. As Heider (1958) pointed out long ago, we are all naive psychologists attempting to understand and explain the causes of our outcomes. While most organizational scientists accept the basic premises of reinforcement theory (Miner, 2003), they appear to be a bit less certain of the role of attribution theory. I believe that this uncertainty can be overcome if attributions are considered within the context of reinforcement theory. More specifically, attributions are simply individuals' beliefs about the causes of their rewards. If one accepts this conceptualization and also accepts the basic premise that behavior is a function of consequences, it becomes clear that people beliefs about the causes of their rewards are integrally related to their behaviors.

Following the chain of reasoning outlined above, virtually all behaviors affected by reinforcement are also affected by people's beliefs about the causes of their reinforcement (i.e., attributions). Thus, expectancies are shaped by attributions and virtually every behavior that is under the control of reinforcement is also influenced by attributions.

This somewhat broad view of the domain of attribution theory may be perceived buy some researchers and theoreticians to stretch the bounds of traditional attribution theory. As initially described by Kelley (1973) and Weiner (1979), attributions are defined as people's causal explanations for the outcomes of their behavior. Over time, the boundaries of these definitions and the applications of the construct have been expanded. Moreover, as attribution theory is applied to organizational contexts, issues have evolved that are not typically considered in psychology. More specifically, organizational contexts shift the focus from individual to dyadic and group behavior. Examples are the papers from this conference by Dasborough and Ashkanasy (follower attributions for leader behavior), Thomson and Martinko (social attribution styles), Douglas and Joseph and Betancourt (the effect of attribution styles on interpersonal conflict), and Laufer (group attributions for joint ventures). All of these papers place attributions into a larger social context than has been typical. Questions such as how groups evaluate and assess organizational performance become much more salient in this context. As a result of these more macro issues, theoretical adaptations have occurred and are represented in many of the papers that make up the chapters of this book. In my own work, I have moved from the more narrow concept of attribution theory to the concept of causal reasoning, which encompasses not only individuals' explanations for the outcomes of specific behaviors, but also their more general beliefs about cause and effect in their environment. Thus, when encountering a new situation, I assume that people generalize the elements of the new situation from past experiences and make attributions regarding the likely cause of outcomes in this new environment. In the hope of not confusing this somewhat broader perspective with more traditional uses of attribution theory, I have labeled these types of processes as causal reasoning. I believe that the notion of causal reasoning is particularly valuable in organizational contexts where we are often talking about strategic behaviors, attempting to anticipate long-term consequences.

KEY ISSUES

Many of the issues we discussed as a group are related to people's assumptions about the nature and importance of attributional processes. As a part of the conference we broke off into discussion groups that centered on key issues. While I will attempt to summarize the content of these discussions, I will also provide my own perspective regarding these issues.

Application

One group of participants addressed the issue of the application of attribution theory. Some of the members of the group had adopted a broad perspective of attribution theory similar to the one described above and saw a wide range of application. They argued that attribution theory could and should be integrated into almost all major organizational behavior topics such as leadership, organizational citizenship, impression management, counterproductive behaviors such as turnover and aggression, conflict management, and strategic planning. Thus, they believed that the explanatory power of the theory in each of these areas could be enhanced by explaining and understanding how attributional processes affected each of these behavioral domains.

There was also some disagreement regarding the broad application of attribution theory. The argument was that attribution processes require deliberate cognitive effort and that many routine organizational behaviors are scripted and do not have a deliberative component. A similar argument was made at the earlier attribution conference by Lord (1995) and also in a paper by Mitchell (1982). Thus, Lord argued that more primitive and implicit as well as cybernetic behaviors couldn't efficiently be controlled by deliberate cognitive processes such as attributions because of time and expediency elements.

The rebuttal to these arguments is that, over time, attributional processes become abstracted and scripted. Just as Piaget had described the evolution of mathematical processes that evolve from concrete operations (i.e., counting on the fingers) to abstract schematic representation (use of numbers and formulas), it can be argued that after experiencing numerous sequences of cause and effect, these experiences are internalized and eventually evolve into schema that cue scripted behaviors that appear almost automatic and routine until some unexpected, negative, or particularly important consequence occurs (Weiner, 1985).

All of the above arguments are plausible. From a theoretical perspective, what matters is whether or not considering attributional processes adds explanatory power to our theories. It does not appear that the differences between these various positions can be resolved through logic and debate. However, as additional empirical work develops, the results could help clarify the extent to which attributions play a useful role in explaining the wide array of behaviors proposed.

Leadership

There were two schools of thought regarding the application of attribution theory to leadership. The first emphasized the process by which individuals perceive that an individual is characterized as a leader. In this context, the use of the term *attribution* is closer to what organizational scientists typically label a perception, rather than the more conventional use of an attribution as a term denoting the cause of an outcome. Pfeffer's (1977) work on the ambiguity of leadership is an example. In this paper he argues that the construct of leadership derives, at least in part, from the need for individuals to perceive control through the beliefs that leadership is a proximal cause of organizational effectiveness. Another example of this approach is the recent work by Gardner and Avolio (1998) that explains how individuals label leaders as charismatic. Although this perspective can be viewed as somewhat inconsistent with more traditional uses of attribution theory, it can also be adapted to a more traditional perspective. More specifically, if one considers "leadership" or "charisma" as attributional explanation for the outcomes of the behavior of an individual (e.g., she achieved this outcome because of her leadership), it now makes sense to analyze the underlying dimensions of this explanation and speculate on how these dimensions will effect the affect and behavior of the attributor. Thus, it would be interesting to see if "leadership" in this context is considered by the attributor as an internal, stable, and controllable characteristic and to see how these attributions and attributional dimensions affect the way the attributor feels and acts toward the leader. This line of research could clearly contribute to both our understanding of leadership and attributional processes.

Another potential contribution of attribution theory to the area of leadership, related to the above example, is concerned with how the attributions that both leaders and members make for each other's behaviors affect leader–member relationships. Earlier research by Mitchell and his colleagues (e.g., Mitchell & Wood, 1980; Wood & Mitchell, 1981), which predicted how leaders would feel and behave toward their members as a function of the attributions leaders make for members' outcomes, reflects this line of research. More recent work by Martinko and his colleagues (e.g., Martinko, 2002; Martinko, Moss, & Douglas, 2003; Thomson & Martinko, Chapter 9, this volume) describes and documents how attribution styles of both leaders and members interact, suggesting that leader and member conflict can be the result of incompatible attribution styles.

Emotion

The relationship between emotion and attributions was a major theme throughout the conference. While the importance of emotions and affective states has only recently become a central topic of organizational behavior (e.g., Dasborough & Ashkanasy, 2002; Pescosolido, 2002), Weiner's work (e.g., 1986) has emphasized the relationship between attributions and emotion for quite some time and many of the current papers reflect the connections between attributions, affect, and behavior. Thus, the relational sequence of thinking (attributions), feeling (affect), and doing (behavior) was a constant theme throughout the conference. The general consensus of the discussion regarding attributions and emotions was that emotions appear to be a significant mediator between attributions and behavior. Given the relative recency of the integration of both the emotion and attributional literatures in the organizational sciences, plus the broad range of organizational behaviors that are influenced by both emotions and attributions, it appears that this is a particularly fruitful area for further research.

Measurement

Another key issue identified during the conference was concerned with the measurement of attributional processes. Traditionally, researchers agreed that it was more appropriate to measure attributional dimensions (e.g., internality and stability) as opposed to attributional explanations (e.g., effort and task difficulty) and the majority of the scales that are used to measure attributions and attributional styles focus on causal dimensions (e.g., Peterson & Villanova, 1988; Russell, 1982). The rationale for this convention is that the underlying dimensions of attributions are believed to be more predictive of emotions and expectancies (and therefore behavior) than are the actual explanations, because the underlying dimensions of explanations are sometimes unclear. Thus, although two individuals may both attribute their poor performance to a lack of ability, one may perceive ability as internal and stable and not expend effort to improve performance while the other may perceive ability as malleable (i.e., unstable) and may increase efforts to become successful. Thus, the caveat has been to not assume the underlying dimensions of attributional explanations. As a result, attributional scales are typically designed to measure attributional dimensions rather than attributional explanations.

Despite this seemingly sound rationale for measuring dimensions rather than explanations, the reliabilities of attributional measures have sometimes been unacceptable, particularly for the internality dimension. I suspect that this is often because respondents think in terms of explanations rather than dimensions. Thus, when someone fails a test, response alternatives focusing on explanations such as effort and ability make intuitive sense to the respondents. On the other hand, asking whether the cause is internal or external and stable or unstable may not make as much intuitive sense to the respondents and their uncertainty undoubtedly affects reliabilities.

Another problem with current procedures for measuring attributions is that not all explanations are amenable to classification along the attributional dimensions that have been identified. For example, failing because of poor strategy does not appear to be easily classified as either internal or external or as a stable or unstable cause. This observation suggests that there may be other dimensions that have not yet been identified.

I also have some concern as to what we classify as an explanation and what we classify as a dimension. Although intentionality and controllability are often classified as dimensions, they could also be considered explanations. Thus, for example, it is conceivable that when a leader experiences a negative outcome because of a member's behavior, she may attribute (i.e., explain) the outcome as the result of the intent of the member and punish the member because the cause of the outcome is internal and controllable. Thus there does not appear to be a clear distinction between attributional explanations and attributional dimensions.

The chapter on workplace aggression addressed some of these issues in that the measurement procedures included assessments of both attributional dimensions and attributional explanations and then analyzed the results of both of these procedures separately. There were essentially no differences in the conclusions using either method.

There appear to be no easy solutions to the measurement issues described above and this reality may have been reflected during the symposium in that, although everyone agreed that measurement was an issue, only one person elected to participate in a discussion group to address measurement issues.

Impression Management

Impression management is another area that would appear to benefit from the integration of attribution theory. By definition, impression management is the management of behaviors to present a desired image (Gardner & Martinko, 1988). If we assume that the purpose of impression management is to maximize the probability that individuals are reinforced by others, creating attributions that result in positive affect that is likely to be followed by reinforcement becomes the goal of impression management behaviors. Again, with a few exceptions (e.g., Martinko, Douglas, Ford, & Gundlach, 2004), there has been little work linking attributions to impression management strategies in either the organizational or psychological literature. This appears to be another area where significant contributions can be made.

Cross-Cultural Issues

There was also significant interest in the differences in attributions and attributional styles across cultures. Numerous studies have documented these differences. Examples include the research of Betancourt and Weiner (1982) and Powers and Wagner (1983), demonstrating that people from some Spanish-speaking cultures tend to make more external attributions than subjects from English-speaking cultures and the work of Morris and Peng (1984) that found that Chinese subjects were more external in their attributions than American subjects. Of particular interest here is not so much the observation that there are differences in attributions associated with culture but the underlying reasons for these differences, which could be related to socioeconomic status, the political realities of the various cultures, and even the structure of language. Thus, for example, if someone says that they lost their car keys in Spanish, the literal translation in English is that the car keys lost themselves. The notion that language and culture may shape attributions and attribution styles is intriguing and a more complete understanding of the underlying behavioral dynamics of these observations would undoubtedly enhance our understanding of both attribution theory and the differences between cultures in the workplace.

CONCLUDING THOUGHTS

Once again, one of the reasons I believe that attribution theory can make such a strong contribution to understanding organizational behavior is its strong links to reinforcement processes. One of the strengths of reinforcement theory is the ability to claim a scientific perspective focusing on observable and measurable behavior. Without getting into debate regarding the scientific validity of cognitive processes, it is clear that one of the most solidly grounded links between observable and measurable behavioral processes and cognitions is attribution theory, which focuses on people's beliefs about their reinforcement processes. The chapters in this book demonstrate the strong links to reinforcement processes but are only a sample of the potential that attribution theory has for understanding behavior in organizations. Hopefully, this book will serve as a stimulus that can help further our understanding of the fundamental dynamics of organizational behavior and the contribution that attribution theory can make to the evolution of our knowledge. It is also hoped that in adapting attribution theory to organizational contexts, attribution theory will be enriched and extended.

REFERENCES

Betancourt, H. & Weiner, B. (1982). Attributions for achievement-related events, expectancy, and sentiments: A study of success and failure in Chile and the United States. *Journal of Cross Cultural Psychology, 13*(3), 362–374.

Dasborough, M., & Ashkanasy, N. (2002). Emotion and attribution of intentionality in leader–member relationships. *The Leadership Quarterly,* 13(5), 615–634.

Gardner, W. L., & Avolio, B. (1998). The charismatic relationship: A dramaturgical perspective. *Academy of Management Review, 33*(1), 32–58.

Gardner, W. L., & Martinko, M. J. (1988). Impression Management in Organizations. *Journal of Management, 14*(2), 321–338.

Heider, F. (1958). *The psychology of interpersonal relations.* New York: Wiley.

Kelley, H. H. (1973). The process of causal attributions. *American Psychologist, 28,* 107–128.

Martinko, M. J. (2003). *Thinking like a winner: A guide to high performance leadership.* Tallahassee, FL.: Gulf Coast.

Martinko, M. J., Douglas, S. C., Ford, R., & Gundlach, M. J. (2004). Dues paying: A theoretical explication and conceptual model. *Journal of Management, 30*(1), 49–69.

Martinko, M. J., Moss, S., & Douglas, S. C. (2003) *The effects of attribution styles on leader-member relations.* Paper presented at the Academy of Management Conference. Seattle, WA.

Miner, J. B. (2003). The rated importance, scientific validity, and practical usefullness of organizational behavior theories: A quantative review. *Academy of Management Learning and Education, 2,* 250–269.

Mitchell, T. R. (1982). Attributions and actions: A note of caution. *Journal of Management, 8,* 65–74.

Mitchell, T. R., & Wood, R. E. (1980). Supervisor's responses to subordinate poor performance: A test of an attributional model. *Organizational Behavior and Human Performance, 25*(1), 123–138.

Morris, M., & Peng, K. (1994). Culture and cause: American and Chinese attributions for social and physical events. *Journal of Personality and Social Psychology, 7*(6), 949–971.

Pescosolido, A. (2002). Emergent leaders as managers of group emotion. *Leadership Quarterly, 13*(5), 583–599.

Peterson, C., & Villanova, P. (1988). An expanded attribution style questionnaire. *Journal of Abnormal Psychology, 97*(1), 87–89.

Pfeffer, J. (1977). The ambiguity of leadership. *Academy of Management Review, 2*(1), 104–112.

Powers, S., & Wagner, M. (1983). Attributions for success and failure of Hispanic and Anglo high school students. *Journal of Instructional Psychology, 10*(4), 171–176.

Russell, D. (1982). The cuasal dimension scale: A measure of how people see causes. *Journal of Personality and Social Psychology, 42,* 1137–1145.

Weiner, B. (1979). A theory of motivation for some classroom experiences. *Journal of Educational Psychology, 71,* 3–25.

Weiner, B. (1985). An attributional theory of achievement motivation and emotion. *Psychological Review, 92,* 548–573.

Wood, R. E., & Mitchell, T. R. (1981). Manager behavior in a social context: The impactof impression management on attributions and disciplinary action. *Organizational Behavior and Human Performance, 28,* 365–378.

INDEX

Attribution Theory in the Organizational Sciences, pages 307–312
Copyright © 2004 by Information Age Publishing